ADVANCES IN SYSTEMS BIOLOGY

ADVANCES IN EXPERIMENTAL MEDICINE AND BIOLOGY

Recent Volumes in this Series

Volume 539
BLADDER DISEASE, Part A and Part B: Research Concepts and Clinical Applications
Edited by Anthony Atala and Debra Slade

Volume 540
OXYGEN TRANSPORT TO TISSUE, VOLUME XXV
Edited by Maureen Thorniley, David K. Harrison, and Philip E. James

Volume 541
FRONTIERS IN CLINICAL NEUROSCIENCE: Neurodegeneration and Neuroprotection
Edited by László Vécsei

Volume 542
QUALITY OF FRESH AND PROCESSED FOODS
Edited by Fereidoon Shahidi, Arthur M. Spanier, Chi-Tang Ho, and Terry Braggins

Volume 543
HYPOXIA: Through the Lifecycle
Edited by Robert C. Roach, Peter D. Wagner, and Peter H. Hackett

Volume 544
PEROXISOMAL DISORDERS AND REGULATION OF GENES
Edited by Frank Roels, Myriam Baes, and Sylvia De Bie

Volume 545
HYPOSPADIAS AND GENITAL DEVELOPMENT
Edited by Laurence S. Baskin

Volume 546
COMPLEMENTARY AND ALTERNATIVE APPROACHES TO BIOMEDICINE
Edited by Edwin L. Cooper and Nobuo Yamaguchi

Volume 547
ADVANCES IN SYSTEMS BIOLOGY
Edited by Lee K. Opresko, Julie M. Gephart, and Michaela B. Mann

Volume 548
RECENT ADVANCES IN EPILEPSY RESEARCH
Edited by Devin K. Binder and Helen E. Scharfman

Volume 549
HOT TOPICS IN INFECTION AND IMMUNITY IN CHILDREN
Edited by Andrew J. Pollard, George H. McCracken, Jr., and Adam Finn

A Continuation Order Plan is available for this series. A continuation order will bring delivery of each new volume immediately upon publication. Volumes are billed only upon actual shipment. For further information please contact the publisher.

ADVANCES IN SYSTEMS BIOLOGY

Edited by

Lee K. Opresko
Pacific Northwest National Laboratory
Richland, Washington

Julie M. Gephart
Pacific Northwest National Laboratory
Richland, Washington

Michaela B. Mann
Pacific Northwest National Laboratory
Richland, Washington

Springer Science+Business Media, LLC

ISSN 0065-2598

ISBN 978-1-4613-4695-1 ISBN 978-1-4419-8861-4 (eBook)
DOI 10.1007/978-1-4419-8861-4

Originally published by Kluwer Academic / Plenum Publishers, New York in 2004
Softcover reprint of the hardcover 1st edition
http://www.wkap.nl/

10 9 8 7 6 5 4 3 2 1

A C.I.P. record for this book is available from the Library of Congress

ACKNOWLEDGMENTS

This proceedings document was prepared by Pacific Northwest National Laboratory Scientific and Technical Information staff members Michaela B. Mann and Julie M. Gephart (editors) and formatted by Rose M. Urbina (document design). The papers were reviewed by Pacific Northwest National Laboratory scientists Fred J. Brockman, Lee K. Opresko, Karin D. Rodland, Thomas C. Squier, and H. Steven Wiley.

Special thanks go to Gwen M. (Wendy) Owen for organizing the Northwest Symposium for Systems Biology and handling all logistics. Pacific Northwest National Laboratory is operated by Battelle Memorial Institute for the U.S. Department of Energy under Contract DE-AC06-76RL01830.

TABLE OF CONTENTS

GENE REGULATORY NETWORKS

1. INTRODUCTION

About the Northwest Symposium for Systems Biology

This publication is the proceedings of the Pacific Northwest National Laboratory (PNNL) inaugural meeting of the Northwest Symposium for Systems Biology, held October 17 and 18, 2002, in Richland, Washington. This is the 40th year in which the laboratory has held an interdisciplinary science symposium to address important biological questions. In years past, the unifying theme was environmental sciences. This year we began a new series of symposia on systems biology. A particular focus of these symposia will be on identifying current breakthrough technologies and their application to important model systems.

PNNL established the Biomolecular Systems Initiative (BSI) to exploit the unique and innovative technologies developed here at the laboratory, especially at the William R. Wiley Environmental Molecular Sciences Laboratory (EMSL). The BSI is a multidisciplinary research program that focuses on the areas of research that will drive biology in the post-genomic era. It combines cutting-edge capabilities for high-throughput proteomics, cell imaging, quantitative biology, and computational biology.

To understand complex biological systems, scientists must acquire detailed knowledge about cell signaling, and about how networks regulate cell functions. This will require an integrated effort across a variety of research disciplines: molecular and cellular biology, biochemistry, physics, mathematics, and information science. The BSI is working to provide opportunities for scientists from different disciplines to gather and discuss cell networks at all scales as well as approaches for understanding the molecular components of these networks.

The theme of this year's symposium was the U.S. Department of Energy's new Genomes to Life (GTL) program. GTL has the eventual goal of a fundamental, comprehensive, and systematic understanding of life. In its initial implementation, GTL focuses on post-genomic approaches to understanding

- Complex Microbial Systems
- Computational Methods
- Molecular Machines: Multiprotein complexes
- Gene Regulatory Networks.

This symposium brought together scientists from the four major areas addressed by the GTL and encouraged them to share their data as well as their broad perspectives on different areas of interest and where they overlap. Approaches were discussed for integrating the wide variety of data needed for a systems-level approach to biology. Both prokaryotic and eukaryotic systems were represented in the symposium, with topics ranging from the analysis of microbial communities to the effect of protein modification on protein function(s) and molecular interaction(s). The breadth of the symposium was stimulating, and the participants were enthused by the different scientific perspectives that could be applied to complex problems in biology. The ability of scientists in different research areas to work productively together will be essential for biology to move past the stage of characterizing individual molecules, such as genes and proteins, and move toward a more comprehensive, integrated view of biology at a whole-systems level.

Lee K. Opresko, Symposium Chair
H. Steven Wiley, Director, Biomolecular Systems Initiative
Pacific Northwest National Laboratory

December 2002

COMPLEX MICROBIAL SYSTEMS

2. HIGH-THROUGHPUT TECHNIQUES FOR ANALYZING COMPLEX BACTERIAL COMMUNITIES

David A. Stahl*

ABSTRACT

A more complete understanding of microbial diversity and the environmental processes they control will require much more than a biotic inventory. It will require a deeper understanding of the basic features of systems organization and inter-population interactions. Communities, not total biomass, control net process rates driving the biogeochemical cycles sustaining the biosphere. Although the general patterns of macro-organismal diversity are relatively well known, spatial and temporal patterns of micro-organismal diversity are essentially unknown. Having tools capable of resolving these patterns is a prerequisite to developing an understanding of the relationship between community structure and function.

This talk discusses conceptual and technical developments that now provide the framework for systematically resolving temporal and spatial patterns of microorganisms and relating those patterns to processes at local and system levels. Of particular emphasis will be ongoing studies using highly parallel analyses with DNA microarrays for intensive monitoring of microbial populations in environmental systems. Although microarray technology is reasonably well established for studies of model organisms in well-defined laboratory settings, the application of this technology to environmental systems of uncharacterized diversity imposes additional demands on implementation; in particular, the requirement for optimized discrimination between target and non-target nucleic acids in complex, and undefined, mixtures. To increase the resolving power (information content) of our DNA microarray format, we are investigating the use of thermal dissociation of hybrids immobilized on individual array elements to resolve target and non-target sequences that differ by a single nucleotide. These studies, combined with specialized algorithms for optimizing the readout of the microarray should serve for informed environmental application. Initial studies have validated the general approach for analyses of sediment systems.

* David A. Stahl, University of Washington, Seattle, WA 98195.

2.1. THE TALK

There are two parts to this discussion. The first part will be more of a personal, philosophical discussion about the need to work with microbial systems holistically, using as an example one piece of a graduate student's research that helps demonstrate the motivation for working with systems holistically. Secondly, I will address some recent technology developments that enable more complete analysis of complex microbial systems. The second part will be mostly nuts and bolts.

One of today's major challenges is the rapid acquisition of DNA sequence information, which far exceeds our ability to assimilate it. Another challenge is to bring that data back to the environment; that is, the need to examine that data in relation to the genome's native environment, which in most cases is the soil, air, and water outside the laboratory. This will be the only manner in which we can resolve questions about the functional (physiological), ecological, and evolutionary factors that structure each genome.

A major advantage of today's newer technologies is that we are better able to leave the laboratory and move into the environment to conduct microbiology research. A report just published by the American Academy for Microbiology entitled *Microbial Ecology in Genomics: A Crossroads of Opportunity* (Stahl and Tiedje, 2002) discusses a current crossroads of opportunity, the intersection of ecological and genomic research, and addresses the challenge of employing genomics technology in the environment. The report embodies at least part of what I will be presenting today; that is, that the environment is the context in which genomes evolved function, and continue to evolve it, and the environment is the only context in which genomes can be fully understood. That report outlines a proposed 10-year study in which techniques, outreach and training, and targeted areas for specific research programs will provide a road map for a structured, rapid integration of genomics with microbial systematics, evolution, and ecology.

Important considerations in the study of microbial ecology are "emergent properties" and context. I define emergent properties as those properties of a system that are not readily predicted by the analysis of system elements in isolation. These properties are one consequence of the hierarchical structure of living systems. Hierarchy is a well-recognized feature of biology. From simple to complex, the order of biological structures as it is currently understood is monomer, polymer, organelle, cell, population (tissue), community (organ), ecosystem, and biosphere. "Findings made at lower levels usually add very little toward solving the problems posed at higher levels" (Mayr, 1982). This observation has been made many times, and is one aspect of the motivation for moving away from the extremes of reductionism in microbiology research and towards working holistically with complex systems. The abstract painter George Braque (1882-1963) said, "I do not believe in things, I believe only in their relationships," to which I would add a corollary from Mayr: "The species concept [is] a relational concept," (Mayr, 1982). Higher complexity systems are built on the much simpler biological systems that evolved 3.5 billion years ago.

We return, then, to the importance of relating the genome to the biological systems in which they evolved, continue to evolve, and function. The technologies we develop to study the genome must be tuned to, and integrated with, the complex environment in which they function. An organism's interaction with the biotic and abiotic features of its environment determines niche. Therefore, niche should map to genome structure and function. Ultimately, niche will provide the context to define a microbial species in more complete terms. Today, the definition of species remains one of the most contentious areas in microbiology.

The molecular framework that I will use to further develop the conceptual and analytical framework for this presentation is embodied in the small ribosomal subunit RNA (SSU rRNA) phylogeny, sometimes referred to as the universal tree of life or the "Big Tree." Inspection of the big tree reveals that multicellular life was a very late addition to the biology of our planet. Most of evolutionary history is microbial. The planet is driven by unicellular organisms, and the biosphere would do quite well without the big stuff.

Steve Giovannoni of Oregon State University raised the point that diversity is a moving target. We really don't fully understand the extent of microbial diversity—it's only been in the last decade or so that we began to develop a more complete catalog of microbial diversity. Even so, much of the recently recognized microbial diversity, many at approximate kingdom-level taxonomic rank, is only a SSU rRNA sequence placed on the big tree. We generally have no understanding about the physiology of a novel organism represented by SSU rRNA sequence alone. This lack of understanding is something that must become part of the analytical landscape as we develop high-throughput techniques; we must not only learn to deal with what we already know, we must also develop technologies that can deal with what we don't know. This is a significant part of the challenge of modern microbiology.

2.2. ENVIRONMENT-BASED RESEARCH: A HOLISTIC APPROACH

The graduate work of Dr. Jennifer Becker, now at the University of Maryland, provides an example of why we need environment-based studies. Jennifer took on a project examining community adaptation, the process by which microbial communities reorganize ("adapt") to develop the capacity to degrade organic pollutants recently introduced into their local environment. Generally, in microbial adaptation research, there is a period of months, sometimes years, in which nothing—in the sense of transformation—happens. Then a transformation is observed, indicating that something has changed in the system. Nothing is really known about the underlying processes, mostly because we simply did not have a technology suited to direct analysis of microbial systems prior to the advent of molecular tools. Without a complete description of microbial systems, there's really no hope for predicting the fate of pollutants, designing optimized treatment systems, or accelerating the adaptation process. Although this is a very complex problem, Jennifer did a remarkable job of constraining alternative mechanisms of adaptation, and in doing so clearly showed the need for new technological formats.

Jennifer's project embodied most of what I call the current technology for using a combination of molecular and analytical chemical techniques to explore the process of adaptation (see Figure 1). She set up large anaerobic reactor systems, mesocosms of several liters, that were developed using either Lake Michigan sediment from an oligotrophic deep water part of the lake or sludge from a wastewater treatment plant. These mesocosms were either left unexposed to pollutants or were exposed to a set of model chlorinated aromatics. Jennifer measured a number of chemical parameters, as outlined in Figure 1. Transformation of one model pollutant, 3-chlorobenzoate (3-CB),

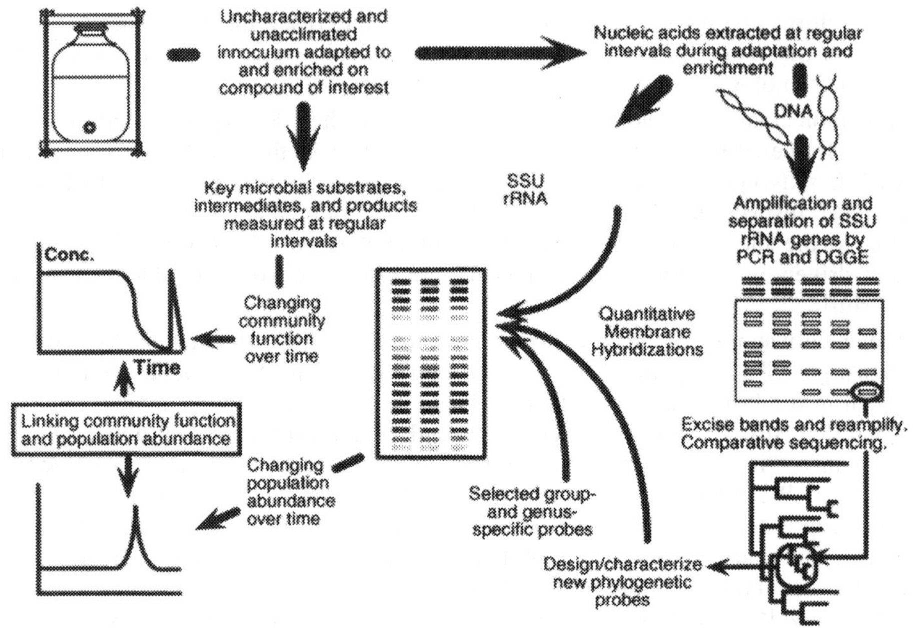

Figure 1. SSU rRNA and ribosomal RNA gene-based approaches used to monitor population changes.

occurred several months after it was added to the mesocosm (Becker, unpublished observations). Subsequent additions were rapidly degraded, demonstrating that the system had adapted to this model pollutant.

The key question is what happened before the rapid onset of pollutant transformation. To address that question, Jennifer used a series of different nucleic acid-based approaches: DNA sequencing and fingerprinting techniques, as well as nucleic acid probes targeting the ribosomal RNA of selected populations. Probe-based quantification completed at the time of these studies used a well-established membrane-hybridization format. This technique remains more or less the state-of-the-art for probe-based quantification. It is rather cumbersome, but offers fairly high-precision measurement of ribosomal RNA abundance. This enabled Jennifer to relate changing population structure to changing process and therefore to better constrain antecedent biological changes occurring during the adaptation process. Since we don't know much about what's out there, Jennifer employed hybridization probes that targeted large assemblages (clades) of microorganisms, rather than individual species. Such probes encompass not only those organisms previously identified, but also those that are undescribed, if they are affiliated with the probe target assemblage. Figure 2 shows an example of two types of hybridization probes targeting the highest taxonomic rank of domain. These probes, about 20 nucleotides in length, provide for comprehensive measurements of community dynamics without the requirement of identifying all contributing populations. However, the format for employing even a collection of general probes (membrane hybridization) restricted the number of samples that could be analyzed.

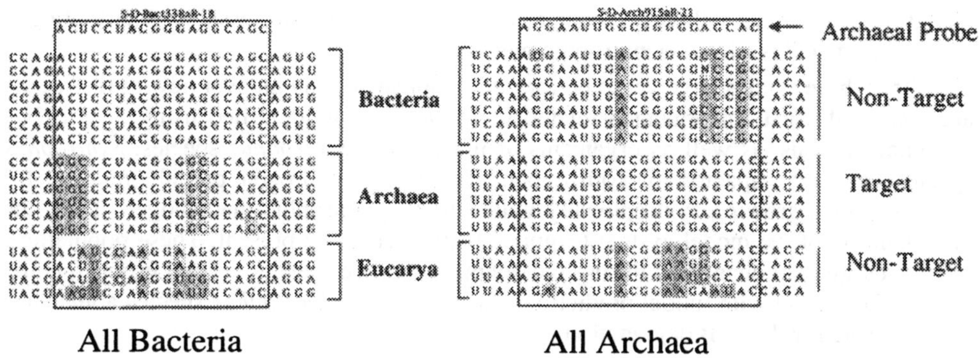

Figure 2. Examples of domain-level probes for bacteria and archaea. Adapted from Stahl and Amann (1991).

A paradigm for how pollutant transformation can occur in an anoxic setting is illustrated in Figure 3. One of the interesting attributes of anaerobic microbial communities is that mineralization of organic compounds tend to be mediated by multiple populations, consortia, such that the metabolic products of one population serve as metabolites for other, metabolically coupled, populations. In this example, 3-CB mineralization requires at least a three-population community. The first member, here *Desulfomonile tiedjei,* grows by halorespiration. It uses 3-CB as an electron sink and in the process removes the chlorine atom to generate benzoate as a metabolite, which is further fermented by an organism that only can oxidize benzoate when coupled to an organism that consumes hydrogen. The thermodynamics are such that anaerobic

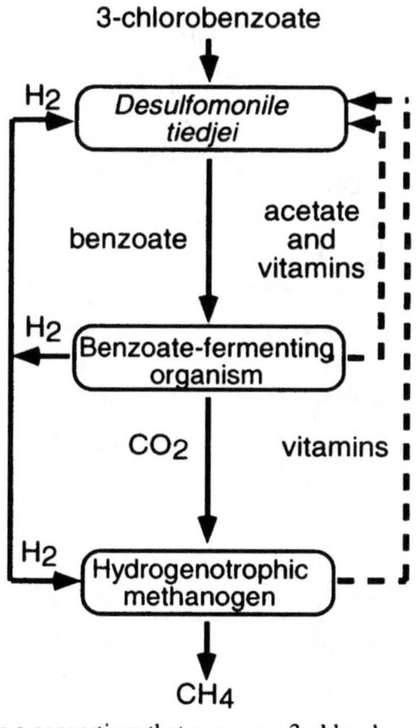

Figure 3. Syntrophic interactions in a consortium that grows on 3-chlorobenzoate. Adapted from Mohn and Tiedje (1992).

oxidation of benzoate will only occur at low hydrogen partial pressures. The hydrogen sink in this model is a hydrogenotrophic methanogen.

Since these studies have yet to be formally published, only an overview of the results are presented. First, an organism related to *D. tiedjei* was identified using SSU rRNA sequence analysis, as well as organisms affiliated with *Syntrophus* species. *Syntrophus* species have the capacity to oxidize benzoate at low hydrogen partial pressure, for example, when growing syntrophically with a hydrogenotrophic methanogen. Thus, the key elements of the model system (Figure 3) appeared to be present. These populations were subsequently quantified during the adaptation process using several group-specific probes. Consistent with model predictions, an increase of *Syntrophus* species abundance was correlated with 3-CB degradation.

It's not always that simple! Jennifer also looked at other model compounds, including 2-chlorophenol (2-CP). Here she inferred that two competing pathways developed in the sediment systems during adaptation (see Figure 4). One pathway involved an immediate reductive dechlorination, much like the 3-CB model system. The other involved an initial carboxylation followed by a dehydroxylation to yield 3-CB, which was further reductively dechlorinated to benzoate. Benzoate is the common intermediate, and could be further mineralized by the combined activities of *Syntrophus*-like and methanogenic populations, as previously described for the 3-CB system (Figure 5). Unlike in the 3-CB system, no delay in the biotransformation of the parent compound was observed in the 2-CP "polluted" system; however, benzoate transiently accumulated during the earlier phases of transformation (Becker et al., 2001). Thus, exposure to different chlorinated aromatic compounds elicited different responses in communities derived from the same uncontaminated environment.

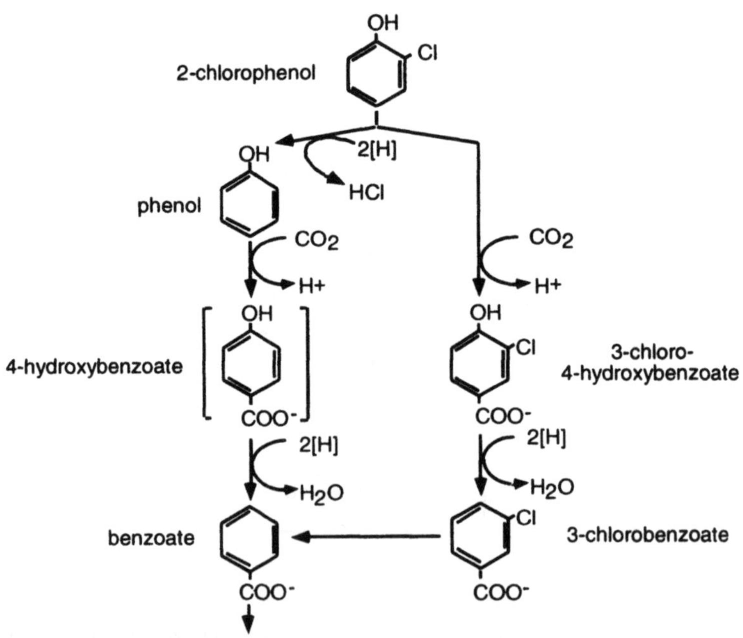

Figure 4. Proposed 2-chlorophenol biodegradation pathways for a sediment microcosm community. Adapted from Becker et al. (1999).

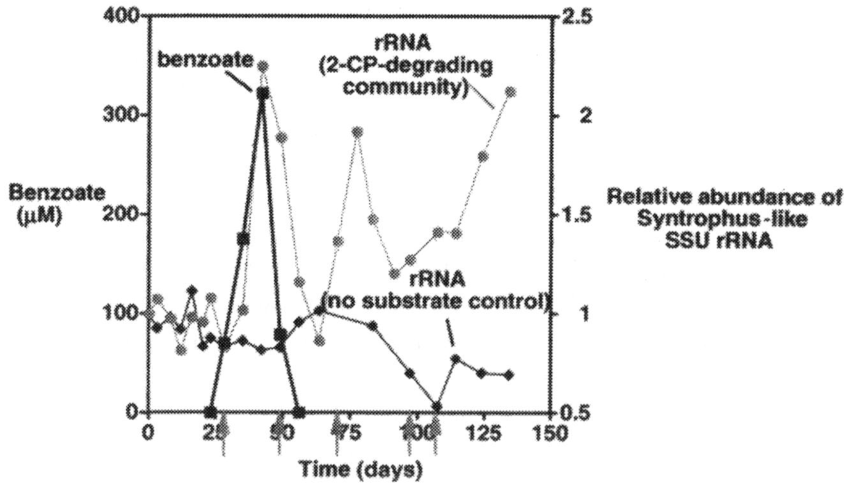

Figure 5. Relationship between abundance of *Syntrophus*-like SSU rRNA and benzoate metabolism in the 2-CP-degradating sediment community. Adapted from Becker et al. (2001).

These studies supported the utility of probe-based analyses of population structure to study the adaptation and biodegradation processes. However, as noted above, the number of samples and population types that could be analyzed was very restricted by the experimental format (membrane hybridization). Jennifer's results also clearly established the need to better resolve the contribution of specific populations during the earlier phases of the adaptation process. However, this will require the application of a much larger set of probes, both general (as used in the described studies) and population-specific. This has been an important motivation for investing in high-throughput technology, as we have done in collaboration with the Biochip group at Argonne National Laboratory for several years.

2.3. TECHNOLOGY DEVELOPMENTS—DNA MICROARRAY TECHNOLOGY

We are now characterizing a DNA microarray format, which is a little different than the standard glass array that many of you are likely most familiar with. This array consists of an array polyacrylamide pads (100 microns on a side and 20 microns deep), which are first deposited on the glass surface. The pads then are impregnated with DNA probes or protein—they have been used in both ways. There are several advantages to using an acrylamide pad rather than direct immobilization on the glass. A key advantage, for our purposes, is achieving a higher probe concentration in one region of the array than is possible using direct immobilization on glass, resulting in improved signal detection.

During initial characterization of the microarray, we simply used our existing collection of probes that were developed using the membrane hybridization technology format I briefly described earlier. These are very general probes that target entire domains and major assemblages of organisms, such as the gram positives and different divisions within the proteobacteria. This is the assemblage that Steve Giovannoni addressed extensively, commonly found in ocean populations. We also took advantage of that array to incorporate much more specific probes that were developed in consideration of the system of study. Thus, the current format of the array that we are developing (Figure 6) is to have four sectors, each of which is divided into four quadrants, each quadrant having

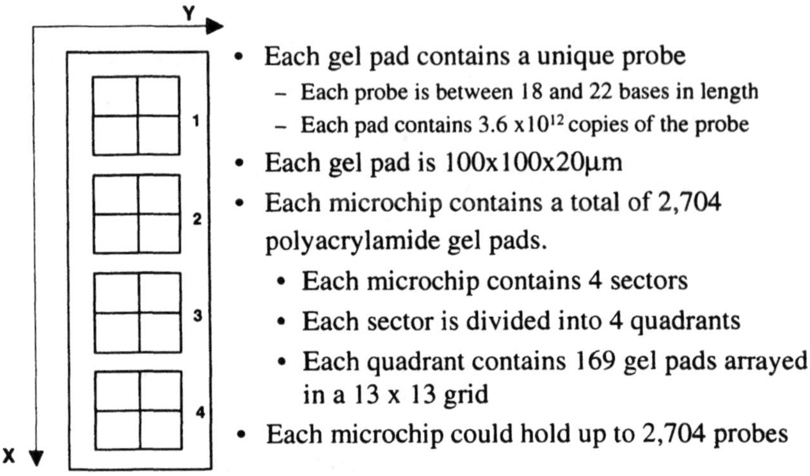

Figure 6. Biochip schematic.

about 169 gel pads. The entire probe number on this particular array is about 2,700. Figure 7 displays the phylogenetic hierarchy used to design the microarray probes.

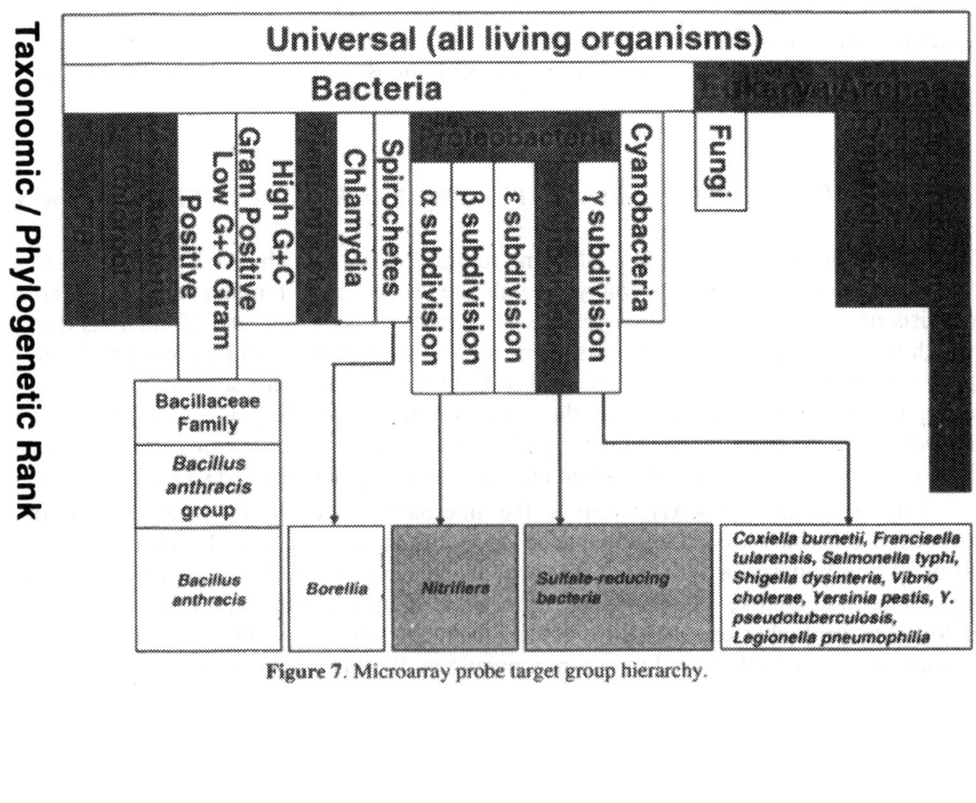

Figure 7. Microarray probe target group hierarchy.

The approach that we have taken is to develop these microarrays both from "top down" and "bottom up." The top-down approach is to incorporate very general probes that target everything we now know about it, but at a very coarse level of resolution. When we work with specific systems, such as systems that are nitrifying or sulfate reducing, we add probes that are system-specific, that target at the approximate taxonomic ranks of genus and species.

The other aspect of the approach we are taking is that currently we are not using polymerase chain reaction (PCR) amplification to interrogate environmental systems. Rather, we extract total nucleic acid from the environment, most of which is composed of the ribosomal rRNAs. This is fragmented using a relatively non-specific free-radical cleavage into fragments of approximately 50-100 nucleotides in length. These fragments are then labeled with a florescent dye and hybridized to the microarray. This method avoids some of the issues of PCR amplification bias. We don't really know how extensive that bias is, but this will be one format where we can begin to directly explore differences between hybridization patterns derived from amplified versus unamplified samples.

Another key attribute of this microarray format is that we don't take a single snapshot following hybridization. Rather, we take advantage of another feature of the system developed at Argonne, in which the microarray is immobilized on a thermal table. This gives us complete control over the temperature of the microarray and allows us to monitor the dissociation of target from probe at increasing temperatures (see Figure 8). We generally generate about 30 intensity readouts for each array element to determine a dissociation profile ("melt curve"). Using that information, we could adjust the wash temperature to achieve optimal readout for each array element (Figure 8, bottom panel). Experimentally this is fairly straightforward if there are two or more mismatches. Bear in mind, however, that we are dealing with a natural system of undefined diversity, not a genome that has been completely sequenced. To interpret hybridization of a microarray to nucleic acid derived from a natural system, we must understand completely what differences in resolution between a perfect match and all possible mismatch variants. Single mismatch variants pose the most difficulty for discrimination.

Dr. Hidetoshi Urakawa, a recent postdoctoral associate in our laboratory, systematically explored the impact of single base-pair mismatches at different positions and base compositions using two model probes for *Staphylococcus* and *Nitrosomonas* species. If we had this type of data set for every probe on an array, we would then be in a good position to optimize discrimination. Of course, with the arrays that we are now working with, we could not build in every possible mismatch for each probe. Our current strategy is to use a more limited collection of well-designed mismatch probes to establish optimal conditions. We have used a simple algorithm to estimate discrimination between probe and non-targets at different wash temperatures. The plot of the discrimination index (DI) value versus temperature tends to be a well-behaved curve, in which the maximum value corresponds to the wash temperature that offers the generally best discrimination between perfect match and single mismatch probes (see Figure 9).

$$DI_{temp} = \underbrace{\left(pm_{temp}/mm_{temp}\right)}_{Discrimination} x \underbrace{\left(pm_{temp} - mm_{temp}\right)}_{\Delta\ Signal\ Intensity}$$

Where DI_{temp} is the maximum at the temperature providing optimum discrimination and signal intensity (Urakawa et al., 2003), pm_{temp} is the average signal intensity of a perfect

- **Determine the T^d for all probe-associated rRNAs. Example (right) melt profiles for perfect match and two different single-mismatch probe duplexes**

- **Display signal intensities at pre-selected temperatures points**

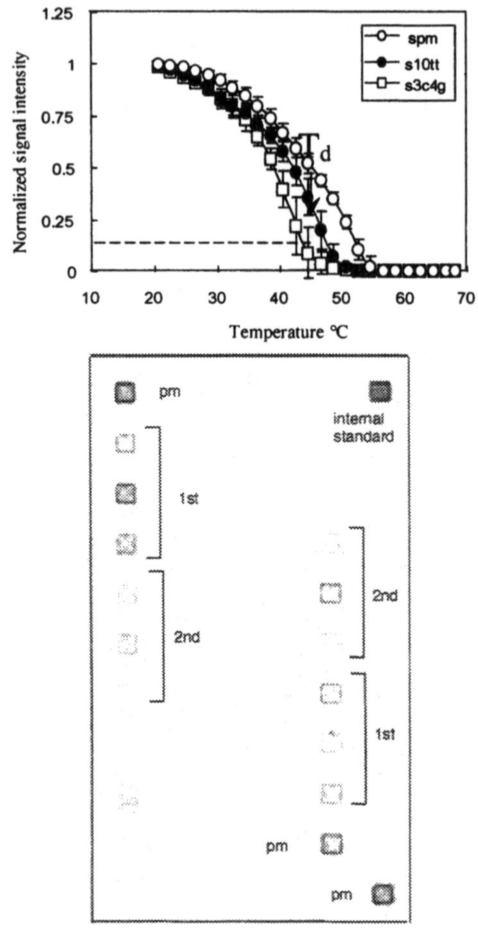

Figure 8. Real-time dissociation and data readout.

match duplex at a specific wash temperature and mm_{temp} is the average signal intensity of mismatched duplexes (excluding those duplexes which have terminal and next to terminal mismatches).

Let me conclude with an example of our first environmental applications of one of these DNA microarrays. This study is ongoing. It is part of a microbial observatory project and long-term ecological research site in Massachusetts at the Plum Island Sound Estuary. The project that we are involved in is looking at nitrification along a salinity gradient, from a fresh-water input through the mid-point of the salinity gradient and out to open water. We know that the potential for nitrification and the populations involved in nitrification vary along this gradient, but we don't really know how that impacts nitrogen processing in the system. One of the larger project goals, then, is to relate the structure of the nitrifier community to actual, and potential, nitrification rates at different points along the salinity gradient. This is primarily the work of Anne Bernhard and Said El Fantroussi, current and past postdoctoral associates in our laboratory at the University of Washington. The discrimination index is shown below.

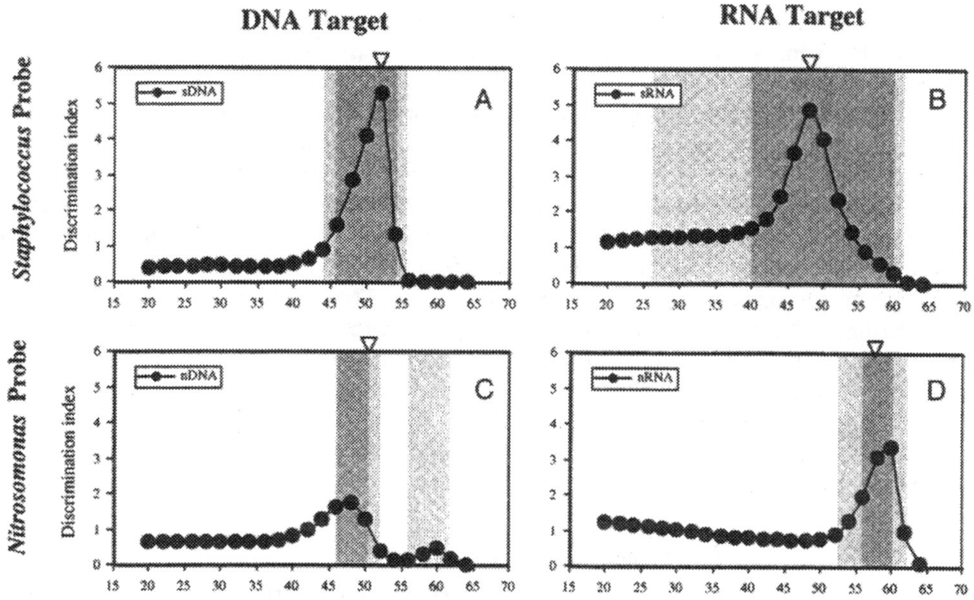

Figure 9. Inferred optimum wash temperatures. Light gray-colored zones indicate temperature intervals allowing for mismatch discrimination as deduced from a neural network (NN) analysis using all data sets (r2>0.7). Dark gray-colored zones are deduced from NN analysis using data sets excluding data from terminal and next to terminal positions (r2>0.9). Adapted from Urakawa et al. (2003).

One of our first analyses used nucleic acid extracted from approximately one gram of sediment material, fragmented and fluorescent dye-labeled (as previously described), for hybridization to our more general microarray (see Figure 10). Notice the region of the array where there is a relatively low signal in the marine part of this estuary and relatively high signals in the fresh-water component. These are all gram-positive targets, which we

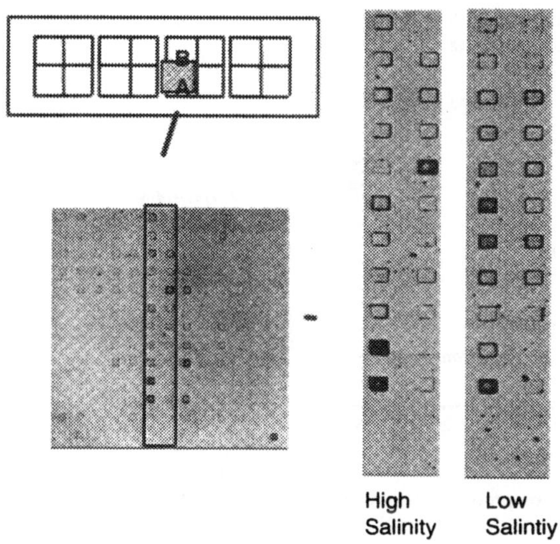

Figure 10. Direct analysis of sediment-derived rRNA. Adapted from El Fantroussi et al. (2003).

know are generally in relatively low abundance in the marine systems. What we see structurally, then, is a clear signal of movement from a fresh-water system to a marine system. The other issue is whether we can we believe the signal. This brings us back again to the utility of melt data for interpreting hybridization specificity. Figure 11 shows the melts for a perfect match and for the environmental RNA. We have good correspondence between these melt profiles, consistent with most of the environmental RNA having a perfect match to the probe target. However, this was not observed for a probe targeting the alpha subdivision of the proteobacteria where the environmental RNA deviated significantly from the perfect-match reference RNA (see Figure 12). This deviation corresponds to a single mismatch deviation. Therefore, most of the target group hybridizing to this probe is a single mismatch non-target. This demonstrates that we can obtain significantly more information by including that melt profile in the interrogation of these microarrays.

We are quite excited about these most recent results because they suggest that this technology will allow us to work directly (without PCR amplification) and at high temporal and spatial resolution in a variety of complex microbial habitats. We anticipate

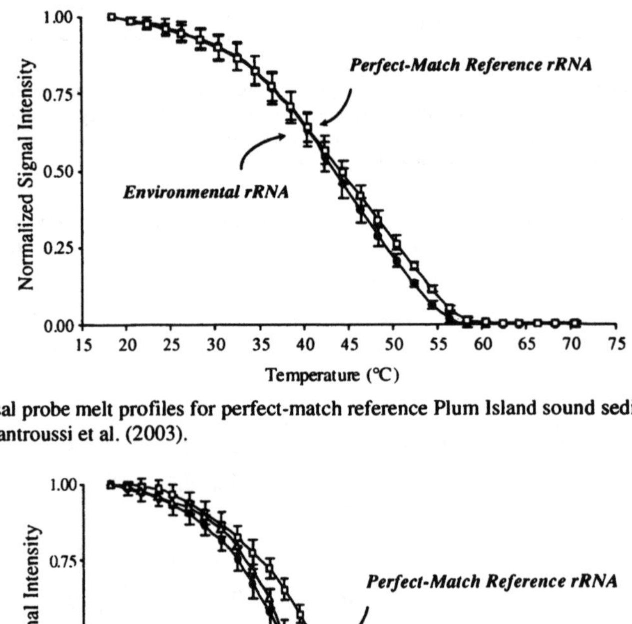

Figure 11. Universal probe melt profiles for perfect-match reference Plum Island sound sediment RNA. Adapted from El Fantroussi et al. (2003).

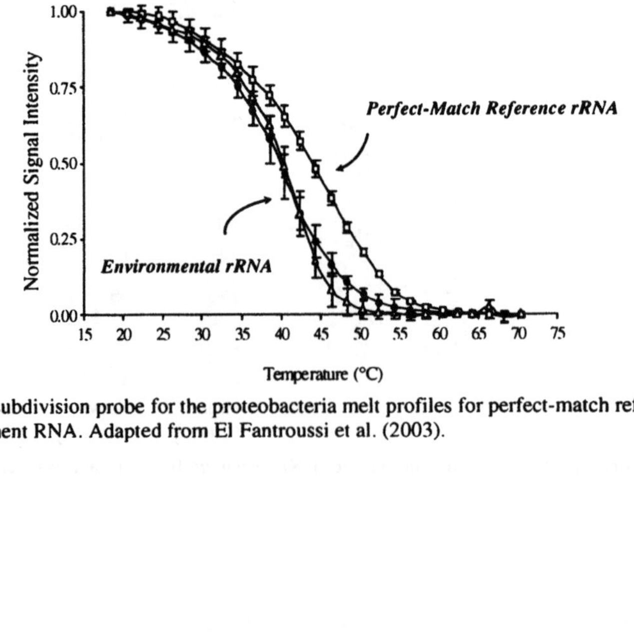

Figure 12. Alpha subdivision probe for the proteobacteria melt profiles for perfect-match reference and Plum Island sound sediment RNA. Adapted from El Fantroussi et al. (2003).

that this and other suitable high-throughput technologies will greatly expand our understanding of microbial ecology, as is needed to control processes in engineered microbiological systems and to predict the behavior of open environmental systems.

2.4. ACKNOWLEDGMENTS

The key microarray work was performed by collaborators at Argonne National Laboratory, Andrei Mirzabekov (leader), Sergei Bavykin, Oleg Alferov, Gennadiy Yershov, Alexander Kukhtin, Victor Barsky, Larry Hill, and Darrell Chandler. Jennifer Becker, University of Maryland, was involved in the adaptation work, and John Kelley, Loyola University, also worked on the microarray. At the University of Washington, Peter Noble, Jim Smoot, Laura Smoot, Said El Fantroussi, and Hideotoshi Urakawa made significant contributions to that work, with help from Hauke Smidt, Wentso Liu (University of Singapore), and Anne Bernhard. Thanks go also to Joany Jackman and Patrick Sullivan, Johns Hopkins/APL, and to John Hobbie and Anne Giblin at Woods Hole Oceanographic Institute. My work is funded by the National Science Foundation, National Aeronautics and Space Administration, the U.S. Department of Energy, and the Defense Advanced Research Projects Agency.

2.5. REFERENCES

Becker, J. G., Stahl, D. A., and Rittmann, B. E., 1999, Dehalogenation and conversion of 2-Chlorophenol to 3-Chlorobenzoate in a methanogenic community: Implications for predicting the environmental fate of chlorinated pollutants, *Appl. Environ. Microbiol.* **65**:5169-5172.

Becker, J. G., Berardesco, G., Rittmann, B. E., and Stahl, D. A., 2001, Successional changes in an evolving chlorophenol-degrading community used to infer relationships between population structure and system-level processes, *Appl. Environ. Microbiol.* **67**:5705-5714.

Fantroussi, S. El, Urakawa, H., Bernhard, A. E., Kelly, J. J., Noble, P. A., and Stahl, D. A., 2003, Direct profiling of environmental microbial populations by thermal dissociation analysis of native ribosomal RNAs hybridized to oligonucleotide microarrays, *Appl. Environ. Microbiol.* In press.

Mayr, E., 1982, *The Growth of Biological Thought*. Harvard University Press, Cambridge, Massachusetts.

Mohn, W. W., and Tiedje, J. M., 1992, Microbial Reductive Dehalogenation, *Microbial Rev.* **56**:482-507.

Stahl, D. A., and Tiedje, J. M., 2002, *Microbial Ecology in Genomics: A Crossroads of Opportunity*. American Academy of Microbiology, Washington, D.C.

Stahl, D. A., and Amann, R., 1991, Development and application of nucleic acid probes in bacterial systematics. *In*: *Sequencing and Hybridization Techniques in Bacterial Systematics*. Stackebrandt, E., and Goodfellow, M. (eds.), John Wiley and Sons, Chichester, England, pp. 205-224.

Urakawa, H., Fantroussi, S. El, Tribou, E. H., Smidt, H., Smoot, J. C., Kelly, J. J., Noble, P. A., and Stahl, D. A., 2003, Optimization of single base-pair mismatch discrimination on oligonucleotide microarrays, *Appl. Environ. Microbiol.* **69**(5):2848-2856.

COMPUTATIONAL METHODS

3. A SYSTEMS APPROACH TO DISCOVERING SIGNALING AND REGULATORY PATHWAYS

—or, how to digest large interaction networks into relevant pieces

Trey Ideker*

ABSTRACT

In the post-genomic era, the first step in any study of protein function is a homology search against the complete genome sequence of the organism of interest. By analogy, if we also wish to elucidate the cadre of signaling and regulatory pathways in the cell, an extremely powerful first step is to construct a complete network of protein-protein and transcriptional interactions and then search through this network to identify critical pathways in a top-down fashion. Like genomic sequence, the molecular interaction network provides a broad foundation for more directed studies to follow. We illustrate this strategy using a large network of 12,232 interactions in yeast. A variety of applications are discussed, including screening the network to identify pathways responsible for gene expression changes observed in galactose-induced cells, and identifying groups of interacting proteins that are essential (by phenotypic assay) for the cellular response to DNA damage.

3.1. INTRODUCTION

In today's post-genomic era, it practically goes without saying that any study of protein function depends on first having a relatively complete genome sequence map of the species of interest. By analogy, if we are interested not just in protein function, but also in how proteins are interconnected within a complex web of signaling and regulatory pathways in the cell, knowing the genome is not quite enough. In addition to the genome, we should also have as our base a comprehensive "interactome"—that is, the network of all protein-protein, protein-DNA, protein-small molecule, and other

* Trey Ideker, Whitehead Institute for Biomedical Research, Massachusetts Institute of Technology, Cambridge, MA 02142-1479.

interactions that drive cell function. Then, just as we might use BLAST to search the genome for particular proteins of interest, novel computational tools will allow us to filter through the interaction network to extract relevant signaling or regulatory pathways of interest.

There are two fundamental approaches for studying this interaction network: (1) directly observing the molecular interactions themselves; and (2) observing the molecular and cellular states induced by the interaction wiring. In terms of the first approach, recent systematic two-hybrid (Ito et al., 2001) and co-immunoprecipitation (Mann, Hendrickson, and Pandey, 2001) studies have resulted in a combined database of 15,000-20,000 protein-protein interactions in yeast. Similarly, a new technology known as ChIP-to-chip analysis allows us to measure protein-DNA interactions at large scale. In this analysis, the first "ChIP" stage uses **Ch**romatin **I**mmuno**P**recipitation to pull down transcription factors of interest and all of the promoters they bind, whereas the second stage identifies promoters bound by each transcription factor by labeling and hybridization against a microarray "chip." Lee et al. (2002) have now performed this procedure systematically for approximately 100 transcription factors in yeast, resulting in about 6000 known protein-DNA interactions. Of course, interactions between proteins or between proteins and DNA are not the only types of interactions mediating signaling and regulatory pathways. Other important interactions occur between proteins and hormones, proteins and drugs, or proteins and metabolites, but cannot yet be measured at large scale.

And as for the second fundamental approach, observing the molecular states induced by the interactions? Certainly, DNA microarrays are now widespread in molecular biology for measuring gene expression changes at large scales. In addition, mass-spectrometry-based approaches are now making it possible to interrogate the abundances and phosphorylation states of many proteins simultaneously. Other molecular states, such as abundance levels for the thousands of intracellular metabolites, cannot yet be measured systematically, although mass spectrometry and NMR promise to revolutionize this area as well.

3.2. INTEGRATING INTERACTIONS AND MOLECULAR STATES

Given databases of interactions and states, there is now a tremendous need for computational models and tools able to integrate these large-scale data within a common modeling framework. One goal of this integration is to search the interaction network to identify particular pathways of interactions that correlate with or explain changes in the molecular state.

For instance, consider the integrated network shown in Figure 1, representing a region of the known interaction network surrounding the process of galactose utilization (GAL) in yeast. A node in this network represents a gene and its protein, whereas a link between nodes (i.e., an edge) represents either a protein-DNA (yellow arrow) or protein-protein (blue line) interaction that has been previously determined by some experimental method. The protein-protein interactions shown here are from the BIND (Bader et al., 2001) or DIP (Xenarios and Eisenberg, 2001) databases, while the protein-DNA interactions are drawn from either TRANSFAC (Wingender et al., 2001) or taken from a recent publication by Lee et al. (2002).

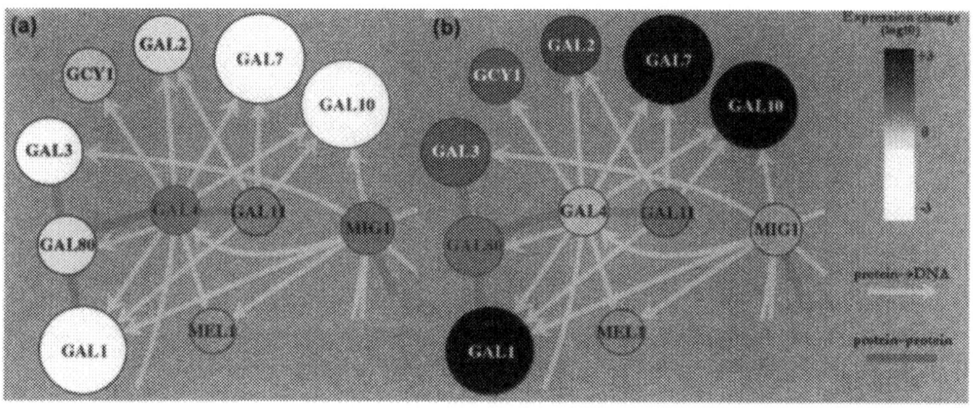

Figure 1. Integrated network representing a region of the known interaction network. Reprinted with permission from Ideker et al. *Science* 292, 929-934 (2001). American Association for the Advancement of Science.

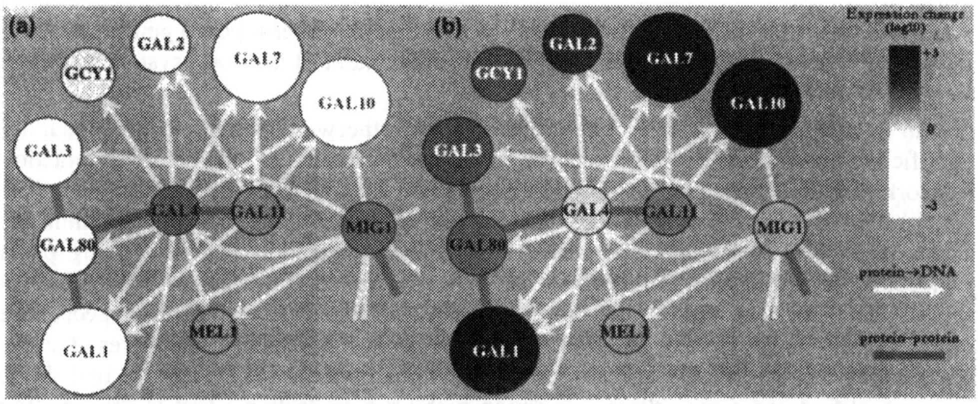

Figure 1. Integrated network representing a region of the known interaction network. Reprinted with permission from Ideker et al. *Science* 292, 929-934 (2001). American Association for the Advancement of Science.

The colors of the nodes represent the states being measured. Figure 1a shows changes in mRNA expression measured in response to a deletion of *GAL4*, whereas the intensities of the other nodes indicate their resulting change in mRNA concentration (Ideker et al., 2001). Background gray represents no change in expression; increasing shades of gray represent increasing levels of mRNA expression; and decreasing shades of gray represent decreasing levels of expression. When *GAL4* is deleted, we see strong decreases in expression of *GAL1*, *7*, and *10*. Importantly, we can begin to explain why we see these changes using interactions present in the underlying network. In this case, the explanation is quite simple: *GAL4* connects directly to *GAL1*, *7*, and *10* through protein-DNA interactions, and it is reasonable to suppose that this is the path by which a *GAL4* deletion evokes these downstream changes.

When we examine different cellular perturbations or biological conditions, the node colors change to reflect these new states. For instance, if we now knock out the *GAL80* gene instead of *GAL4*, the colors reveal a marked increase in *GAL1*, *7*, and *10* (Figure 1b). In this case, a path of length 2 connects *GAL80* to these downstream expression changes: *GAL80* connects to *GAL4* through a protein-protein interaction, while *GAL4* connects to *GAL1*, *7*, and *10* through a series of protein-DNA interactions. In fact, this interaction path turns out to be biologically correct: (Lohr et al., 1995) *GAL80* encodes a repressor protein, which binds to *GAL4* through a protein-protein interaction and keeps it from activating *GAL1*, *7*, and *10*. When GAL80 is knocked out, this protein-protein interaction no longer occurs, and *GAL4* is free to transcribe the GAL genes at a high level.

3.3. AUTOMATICALLY EXTRACTING INTERACTION PATHWAYS FROM THE NETWORK

The galactose-related genes and interactions account for just a small piece of the full yeast molecular interaction network. The full network is actually quite large: recall that the public databases currently contain approximately 20,000 protein-protein and protein-DNA interactions for yeast. In such a large network, we can no longer use a quick visual assessment to pull out putative pathways to explain superimposed gene expression changes. However, the basic ideas illustrated in the context of the GAL system extend to the general case.

In general, when some gene is deleted or otherwise perturbed, the resulting significant expression changes will be distributed about the molecular interaction network. Some of these expression changes may in fact be transmitted from the initial perturbation through a pathway or subnetwork of interactions contained within the network. At a high level, we would like to "connect the dots" by identifying paths connecting perturbed to affected genes. Because of the large number of false positives and negatives in both the interaction and expression data sets, we do not expect these paths to be present or relevant for all gene expression changes. However, for the interactions that are present and transmitting a signal, we should be able to find them. Once identified, we define these interaction pathways as "active"; that is, transmitting expression changes from one gene to another in a particular perturbation or condition. Of course, these "active pathway" hypotheses are only predictions—they must be verified or rejected by directed biochemical assays—but they can be generated automatically.

To search for these pathways and pull them out systematically, we first need a mathematical definition of what it means for a pathway to be active [details of this approach have been previously reported elsewhere (Ideker et al., 2002)]. Consider a network consisting of four proteins A, B, C, and D, as shown in Figure 2. Proteins A and B connect to each other through a protein-protein interaction; proteins B and D regulate C through protein-DNA interactions. Now assume that we have observed gene expression changes over four conditions (rows in Figure 2). We are interested not in the ratio of gene expression, but in the *significance of gene expression change*. Whether the expression ratio goes up or down is irrelevant for the purposes of finding pathways—we are simply looking for regions of change.

To indicate significance of expression, we use an error model and an associated statistical test that assigns z-scores to each expression change in each condition (Ideker et al., 2000). Briefly, this method works as follows: if there is no significant expression

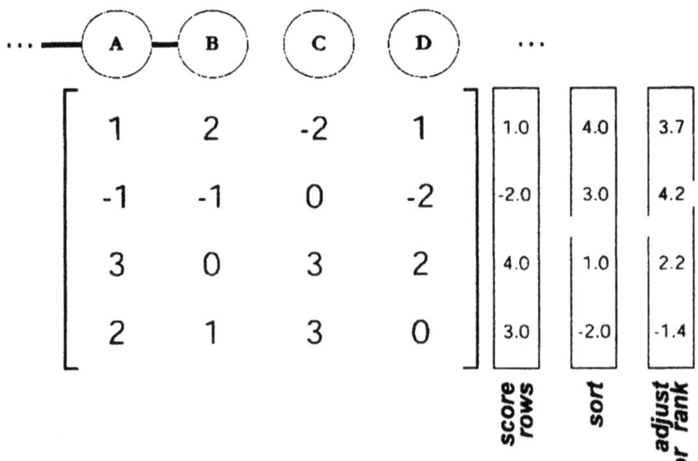

Figure 2. Example interaction path with expression data over four conditions.

change for a gene in a condition, then the z-score follows a standard normal distribution (with mean 0 and standard deviation 1). If there is significant expression change for a gene in a condition, its z-score should be significantly higher than expected by this standard normal distribution. The higher the z-score, the more surprising the gene expression change, whether the gene is induced or repressed. For example, out of all four genes shown in Figure 2, we are most confident that gene B has changed in expression in condition 1. We are somewhat less confident that the levels for A or D have changed in this condition, and we are fairly sure that the level of C has not changed.

Once we have computed z-scores using the error model, it is straightforward to score pathway activity by adding together the z-scores of all nodes in the pathway. If no genes are differentially expressed, this sum will itself follow a standard normal. Otherwise, the sum is significantly higher. For example, to score the pathway ABCD in condition 1, we compute the sum $1+2-2+1 = 2$ and then divide by the square root of the number of nodes (to normalize the z-score back down to standard deviation 1), resulting in an aggregate "pathway activity" score of $2/\sqrt{4} = 1$. Scoring a pathway over multiple conditions is more complex and is explained in full in Ideker et al. (2002).

Scoring a pathway is only half the problem. Given this scoring system, how do we find the absolute highest scoring pathways in the entire network of 20,000 protein-protein and protein-DNA interactions? This problem can be shown to be NP complete, which means that an exact solution is not obtainable in polynomial time. Instead of solving it exactly, we use an approximation algorithm based on simulated annealing. This algorithm finds, if not the single highest-scoring pathway, a collection of several relatively high-scoring "active" pathways. To search for active pathways using simulated annealing, we take the full molecular interaction network (of ~25,000 interactions among 6000 nodes in the case of yeast) and randomly choose several pathways as initial seeds. Then, over a number of iterations, we add/subtract nodes to each pathway in an attempt to improve its score. If the score increases, we keep the change, whereas if the score decreases, we discard the change with a certain probability dictated by annealing temperature. Given enough iterations, the score starts out low and gradually improves until it stabilizes. In this way, the annealing algorithm is guaranteed to produce pathways that have at least a local optimum in score.

3.4. SCREENING FOR ACTIVE PATHWAYS RESPONDING TO GALACTOSE-GENE PERTURBATIONS

Now let's use this automated pathway search procedure to investigate a specific biological problem of interest. In a proof-of-principle application, we recently screened the yeast interaction network to find pathways active under different perturbations to the galactose utilization network in yeast (Ideker et al., 2002). Seven perturbations were performed, by first generating gene knock-outs of *GAL1*, *2*, *5*, *6*, *7*, *10*, and *80* in separate strains, then measuring the corresponding cellular responses with a whole genome mRNA expression profile.

We ran the automated pathway search procedure to identify which pathways from the yeast interaction network were most activated by these perturbations. Five high-scoring pathways were identified and are shown in Figure 3a. As in Figure 1, a line

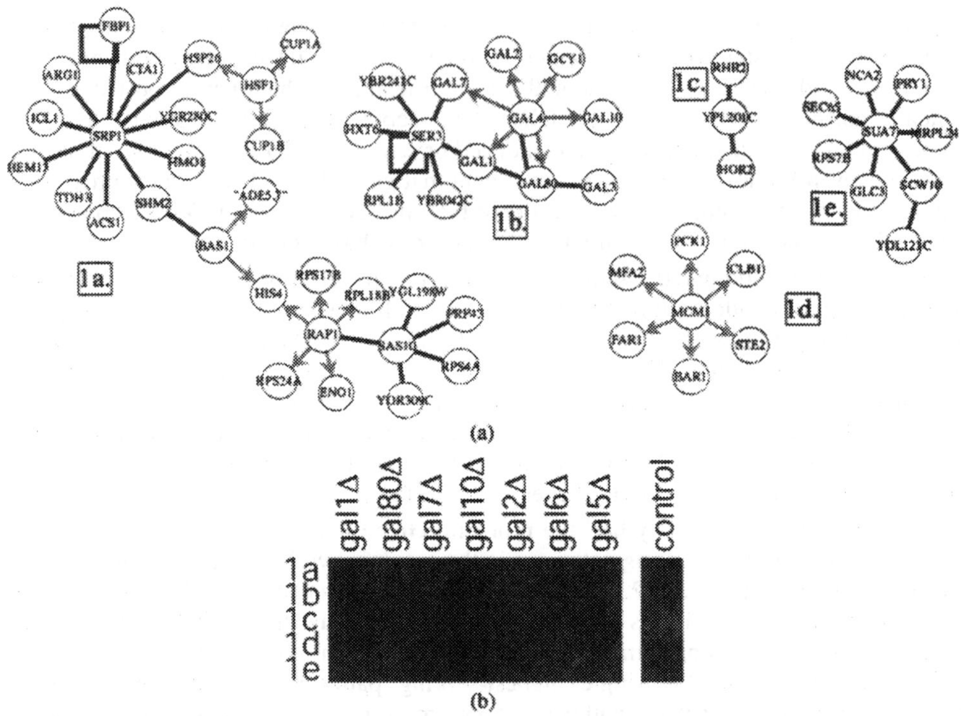

Figure 3. High-scoring pathways. Reprinted with permission from Ideker et al. (2001).

represents a protein-protein interaction and an arrow represents a protein-DNA interaction: all of these interactions are derived and filtered from the whole molecular interaction network.

Figure 3b indicates the particular conditions (columns) activating each of the five pathways (rows). For instance, pathway 1a is active under the *GAL1*, *8*, *7*, *10*, and *2* perturbation experiments, but not under the *GAL7* or *5* perturbation experiments. Likewise, pathway 1b is activated by a *GAL80* perturbation only. Using Figure 3b, we can compare different pathways on the basis of the expression experiments which activate them. For instance, note that pathway 1a and 1c have an identical perturbation profile, which is very different from that of pathway 1b.

The five active pathways represent a combination of known and unknown regulatory processes in yeast. As a "positive control," pathway 1b contains much of the GAL module shown in Figure 1, including the *GAL4* central transcriptional activator and the *GAL80* transcriptional repressor. Given that we are directly perturbing many of the genes in this pathway, we expect it to be active.

Other active pathways represent new discoveries. These provide testable hypotheses for the underlying regulatory and signaling interactions responsible for the observed expression changes. It was not known, for instance, that MCM1 and its downstream regulated genes were involved in the galactose response.

We are currently in the process of applying this approach to a variety of other pathways and expression data sets. One exciting implication for this technology is in the area of drug development. Many drugs are well characterized in terms of what proteins and pathways are being targeted, but not in terms of their possible toxicological side effects. The problem, therefore, is not to discover new drug targets, but to reveal additional pathways that may be affected by the drugs. Here, the limiting factor is obtaining a molecular interaction network relevant to humans. As large interaction networks are determined for key human cell lines—for example, hepatocytes and cancer cells—such an analysis will become possible.

3.5. PATHWAYS RESPONDING TO DNA DAMAGE AS REVEALED BY HIGH-THROUGHPUT PHENOTYPIC ASSAYS

Another method of filtering the molecular interaction network to identify biologically relevant pathways is to use deletion phenotypes. In recent work performed in collaboration with Leona Samson's laboratory (Begley et al., 2002), such an approach was used to map genes and pathways required for the cellular response to DNA damage. For each gene-knockout strain in yeast (libraries of all single gene-knockout strains are now publicly available), we tested whether the strain was able to grow in the presence of MMS, a powerful DNA-damaging agent. Wild-type cells can, in fact, grow under a moderate concentration of MMS, but many gene-knockout mutants either grow slowly or not at all under these conditions.

How do these "MMS-sensitive genes" map onto the protein-protein and protein-DNA interaction network? Figure 4 shows a sampling of interaction pathways containing significant numbers of MMS-sensitive proteins, as determined by the automated pathway screen described in Section 3. In the figure, a node is colored green if deletion of that gene results in slow growth or death in the presence of MMS; red if the deletion has no effect for growth in MMS; and gray if the node has not yet been tested by phenotypic growth assay. Of the gene knockouts tested so far, approximately 400 of them were MMS-sensitive. Using the automated screen for pathways, we were able to associate 100 of these with an "active pathway" having many other MMS-sensitive nodes in close proximity (75 of these appear as green nodes in Figure 4, while the remaining 25 were organized into several pathways not shown in the figure). One interesting observation is that MMS-sensitive nodes may be grouped in a single connected pathway even if several non-sensitive (or non-tested) nodes are required to do so. For instance, to include *MKC7*, *RRP6*, *GIS3*, and *CIN8* in the pathway shown in the upper-left-hand corner of Figure 4, it was necessary to also include YLR453C, which was not tested by phenotypic assay but is included because of its "MMS-sensitive" network neighborhood.

3.6. SUMMARY

A good metaphor for the pathway screening approaches discussed here is that of an information processor, or "black box," as shown in Figure 5. We pour into this black box, on the one hand, all of the molecular interactions previously determined for our organism of interest. On the other hand, we pour in molecular states measured in response to perturbations of a cellular process or biological response of interest. Here, we have used a network of approximately 25,000 protein-protein and protein-DNA interactions in yeast, with state changes measured either at the level of gene expression (Section 4) or growth phenotype (Section 5). After running the "active pathways" algorithm, the black box

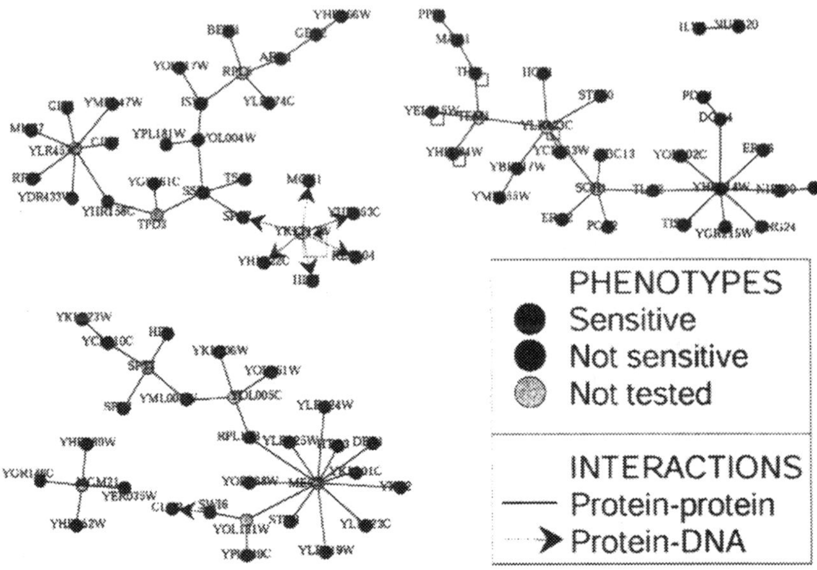

Figure 4. Interaction pathways containing significant numbers of MMS-sensitive proteins. Reprinted with permission from Begley et al. (2002).

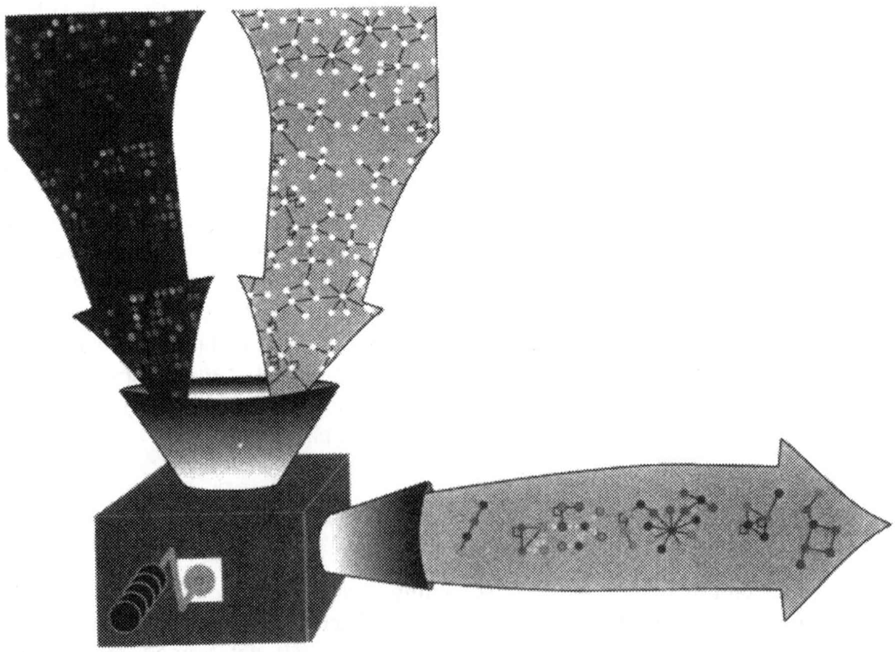

Figure 5. Information processor, or "black box." Reprinted with permission from Ideker and Lauffenburger, *Trends in Microbiology* (in press).

outputs a number of different interaction pathways that appear to be specifically involved in the observed state changes. These pathways are hypotheses that may then be taken into the lab and verified or rejected through directed biochemical or genetic assays.

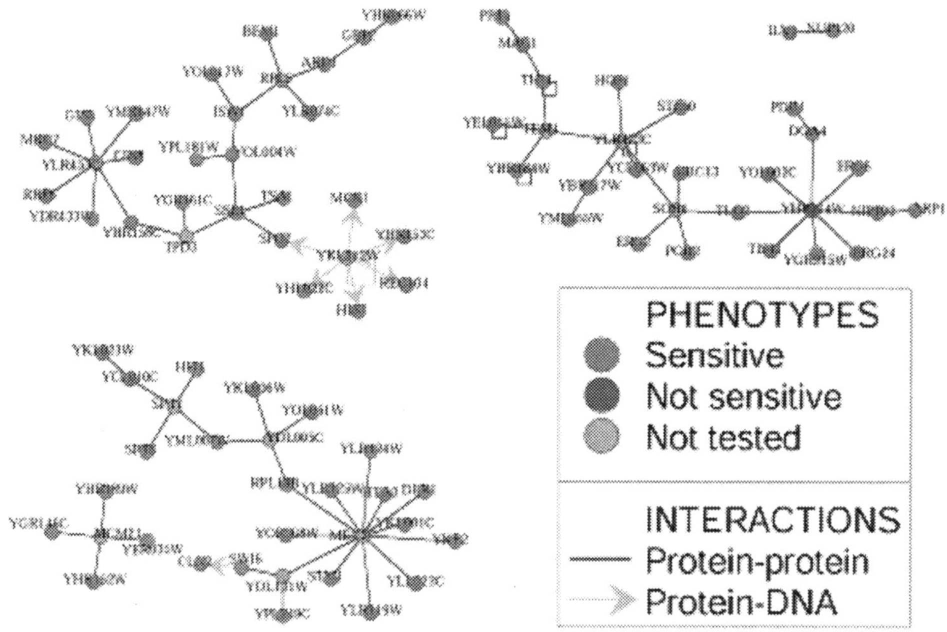

Figure 4. Interaction pathways containing significant numbers of MMS-sensitive proteins. Reprinted with permission from Begley et al. (2002).

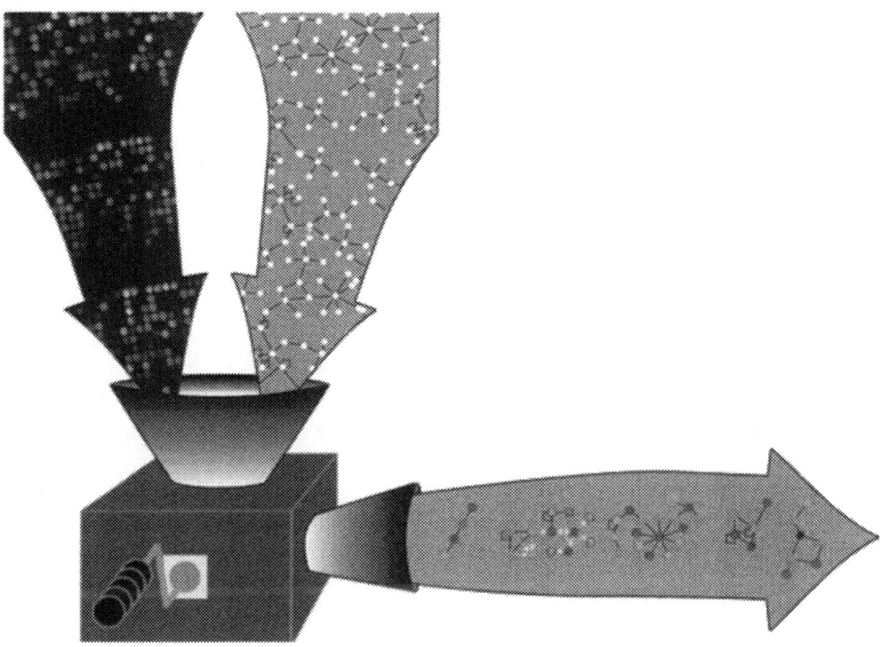

Figure 5. Information processor, or "black box." Reprinted with permission from Ideker and Lauffenburger, *Trends in Microbiology* (in press).

Figures 1, 3, and 4 are based on screen shots taken from a software package called Cytoscape, available as Open Source software from http://www.cytoscape.org as a platform-independent Java application. Cytoscape involves two main components: (1) a core platform for visualizing and manipulating large molecular interaction networks, and (2) an extensible plug-in architecture for writing algorithms and analyses that compute on these networks. The core contains all the routine graphical manipulation, visualization, and information management tasks for large networks: for instance, "How do we lay out these networks in two and three dimensions? Can we link these networks to underlying databases providing annotations for each gene, protein, and interaction"? Plug-ins further extend the basic functionality provided by the core—one such example is the Active Pathway finder discussed in Section 3. Cytoscape is a joint-development project with the Institute for Systems Biology in Seattle.

3.7. ACKNOWLEDGMENTS

This work would not have been possible without the aid of Owen Ozier, Scott McCuine, and other members of the Ideker laboratory; Rick Young (for advance access to data on protein-DNA interactions); and Leona Samson (for her work with the DNA-damage response). Many thanks are also owed to Paul Shannon, Andrew Markiel, and Benno Schwikowski, all at the Institute for Systems Biology, who are co-developers of the Cytoscape software. This work was supported by a grant from Pfizer.

3.8. REFERENCES

Bader, G. D., Donaldson, I., Wolting, C., Ouellette, B. F. F., Pawson, T., and Hogue, C. W. V., 2001, BIND-The biomolecular interaction network database, *Nucleic Acids Res.* **29**:242-245.

Begley, T. J., Rosenbach, A. S., Ideker, T., and Samson, L. D., 2002, Damage Recovery Pathways In Saccharomyces cerevisiae Revealed by Genomic Phenotyping and Interactome Mapping, *Mol. Cancer Res.* **1**:103-112.

Ideker, T., Thorsson, V., Ranish, J. A., Christmas, R., Buhler, J., Eng, J. K., Bumgarner, R., Goodlett, D. R., Aebersold, R., and Hood, L., 2001, Integrated genomic and proteomic analysis of a systematically perturbed metabolic network, *Science* **292**:929-934.

Ideker, T., Ozier, O., Schwikowski, B., and Siegel, A. F., 2002, Discovering regulatory and signaling circuits in molecular interaction networks, *Bioinformatics* **18 Suppl 1**:S233-240.

Ideker, T., Thorsson, V., Siegel, A. F., and Hood, L., 2000, Testing for differentially-expressed genes by maximum likelihood analysis of microarray data, *J. Comput. Biol.* **7**:805-817.

Ito, T., Chiba, T., and Yoshida, M., 2001, Exploring the protein interactome using comprehensive two-hybrid projects, *Trends Biotechnol.* **19**:S23-S27.

Lee, T. I., Rinaldi, N. J., Robert, F., Odom, D. T., Bar-Joseph, Z., Gerber, G. K., Hannett, N. M., Harbison, C. T., Thompson, C. M., Simon, I., Zeitlinger, J., Jennings, E. G., Murray, H. L., Gordon, D. B., Ren, B., Wyrick, J. J., Tagne, J. B., Volkert, T. L., Fraenkel, E., Gifford, D. K., and Young, R. A., 2002, Transcriptional regulatory networks in Saccharomyces cerevisiae, *Science* **298**:799-804.

Lohr, D., Venkov, P., and Zlatanova, J., 1995, Transcriptional regulation in the yeast GAL gene family: a complex genetic network, *FASEB* J **9**:777-787.

Mann, M., Hendrickson, R., and Pandey, A., 2001, Analysis of proteins and proteomes by mass spectrometry, *Annu. Rev. Biochem.* **70**:437-473.

Wingender, E., Chen, X., Fricke, E., Geffers, R., Hehl, R., Liebich, I., Krull, M., Matys, V., Michael, H., Ohnhauser, R., Pruss, M., Schacherer, F., Thiele, S., and Urbach, S., 2001, The TRANSFAC system on gene expression regulation, *Nucleic Acids Res.* **29**:281-283.

Xenarios, I., and Eisenberg, D., 2001, Protein interaction databases, *Curr. Opin. Biotechnol.* **12**:334-339.

4. GENOME FUNCTION—A VIRUS-WORLD VIEW

John Yin*

ABSTRACT

By studying viruses one may begin to understand how static genomes can define dynamic processes of development. This talk will describe some of the approaches we are taking, using computer simulations and laboratory experiments, to account for the many molecular-level processes and interactions that occur when a common bacterium, *E. coli*, is infected by one of its viruses, phage T7. We accounted for processes of phage genome entry, transcription, translation, and DNA replication, including protein-DNA and protein-protein regulatory interactions, and we predicted the dynamics of phage progeny formation. The simulations have enabled us to identify limiting host-cell resources in phage growth, discover novel anti-viral strategies, and suggest frameworks for mining data from global mRNA and protein studies.

4.1. INTRODUCTION

The study of viruses has been primarily driven by the need to understand the causal agents of diseases such as cancer, the common cold, or acquired immunodeficiency syndrome (AIDS). Recently, however, we have increasingly been motivated to view viruses as model genomic systems. From such a perspective, virus genomes encode some of the shortest and most efficient developmental programs. This talk will describe some of the approaches we are taking with the very simplest viruses—the ones that infect bacteria, the so-called "bacteriophages." Our model system has been bacteriophage T7, shown in Figure 1. It is about 60 nanometers in diameter, carries a double-stranded DNA genome 40 kilobases long, encoding 59 genes. When infected by phage T7, an *E. coli* cell at 30°C produces about 100 progeny in a half-hour growth cycle.

The overall growth cycle that we are interested in is shown schematically in Figure 2. The phage lands on its host, which is actually about 10 to 100 times larger than the phage (the figure is not drawn to scale) and starts sending in the linear DNA molecule. It takes about 8 minutes of the half-hour growth cycle for the entire T7 DNA

* John Yin, University of Wisconsin-Madison, Madison, WI 53706.

Figure 1. Electron micrograph of bacteriophage T7, which was used as a model for simulating viral systems. (Image courtesy of F.W. Studier.)

molecule to get into the host. This process is mediated by transcription, initially by use of the host RNA polymerase (RNAP), and later by the phage's own T7 RNAP. Many complex but biochemically and kinetically well-studied steps occur between the initial entry of the genome and the production of phage progeny. Eventually the host cell releases the phage progeny, which go on to find new hosts to repeat the process.

We direct our attention here to the intracellular growth cycle. In particular, we focus on quantifying the dynamics of information flow from the phage DNA to the production of phage progeny.

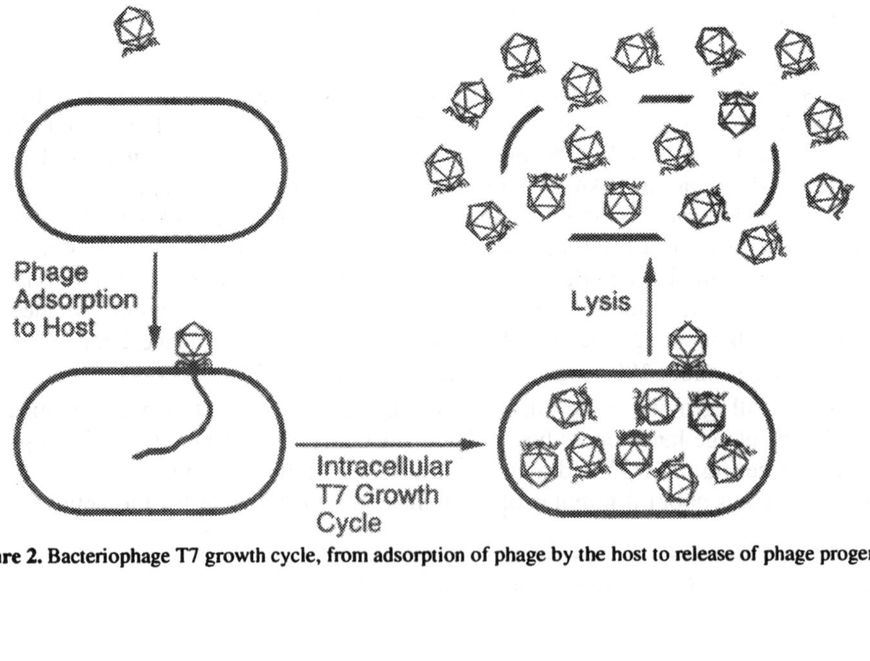

Figure 2. Bacteriophage T7 growth cycle, from adsorption of phage by the host to release of phage progeny.

4.2. THE T7 DEVELOPMENT PROCESS

The full 40 kilobase genome of bacteriophage T7 was sequenced in 1983. It is really quite remarkable that this static linear strand of DNA can encode a complex, dynamic, non-linear developmental process. For a brief look at this process, see Figure 3. In the first phase of the infection cycle, the phage lands on its host and starts sending in DNA. The *E. coli* RNAP binds to its promoter and transcribes early T7 genes. At the same time, the transcription process pulls the T7 DNA into the cell. During this process, the gene for the all-important T7 RNAP is transcribed and translated. The T7 RNAP then recognizes and binds its own promoters, further facilitating transcription-mediated entry of the phage DNA. The phage is quite devious in that it does not allow host resources to be wasted on host processes. It achieves this by synthesizing components that inactivate the host RNAP. In the second stage of the infection cycle, T7 RNAP transcribes a variety of

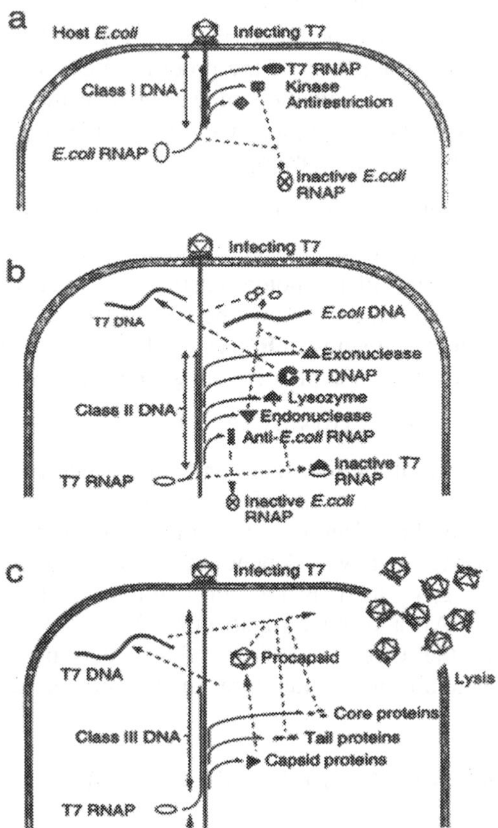

Figure 3. Detail of intracellular phage T7 infection cycle. "a" illustrates phase 1, during which the host RNAP binds to its promoter and transcribes T7 genes, and transcription-mediated entry of the phage DNA occurs. "b" illustrates phase 2, during which T7 RNAP transcribes functions necessary for DNA metabolism and synthesis of progeny genomes. "c" illustrates phase 3, in which phage DNA is packaged into procapsids, producing phage progeny that are released upon lysis of the cell membrane.

different functions that are necessary for DNA metabolism, including the T7 DNA polymerase (DNAP), which plays a central role in synthesizing the progeny genomes. There are also feedback functions and protein-protein interactions that inactivate both the host as well as the T7 RNA polymerase, or at least down-regulate them. In the third and final stage, various structural proteins are made, procapsids are assembled, and the phage DNA is packaged into the procapsids. This is an abbreviated description of what you might find in a textbook on the growth process.

A less user-friendly way of considering the T7 functions is in tabular form (Table 1). From a literature search of T7 spanning the last 40 years one finds a variety of parameters that characterize various T7 functions, such as the elongation rate of the T7 RNAP or the *E. coli* polymerase, or strengths of protein-protein interactions. One also gets other data that relate to resources of the host cell that are essential for T7 growth, such as pool sizes for ribosomes, amino acids, or nucleoside triphosphates. At first glance, Table 1 may not seem very illuminating, but it is a goldmine for someone who does modeling. Table 1 puts numbers to the narrative that accompanies Figure 3.

As engineers, we seek to combine the narrative of Figure 3 and the values in Table 1 to create a mathematical representation of the process. We write conservation equations to account for the synthesis and depletion of each phage mRNA, protein, DNA, and intermediate in the process. For example, in the case of gene product 1, we write an equation for the kinetics of mRNA1 transcription:

$$d(\text{mRNA1}) / dt = \text{rate of growth of mRNA1} - \text{rate of depletion of mRNA1}$$

The magnitude of the first term on the right-hand side, which produces mRNA1, will depend on the relative strength of promoters upstream of gene 1, the elongation rate of the host RNA polymerase, and the length of the mRNA1 transcript. The second term on the right-hand side, which depletes mRNA1, will depend on the rate of mRNA degradation. Likewise, we write an expression for gene product 1(gp1):

$$d(\text{gp1})/dt = \text{rate of growth of gp1} - \text{rate of depletion of gp1}$$

The magnitude of the first term on the right-hand side, which produces gp1, will depend on the level of mRNA1, the spacing of ribosomes on the message, the ribosome elongation rate, and the number of amino acids in gp1, which encodes the T7 RNAP. The second term, which depletes gp1, will depend on the rate of gp1 degradation as well as reactions that consume gp1, such as the formation of gp1-gp3.5, a protein-protein complex between T7 RNAP and the T7 lysozyme. When we create a model, we write such equations for all the essential DNA, mRNA, and protein species, set initial conditions, input parameters from the literature, and numerically integrate the equations.

4.3. FROM EQUATIONS TO INTEGRATED DYNAMICS

By integrating the equations we get a trajectory of intracellular concentration as a function of time post-infection for each component, including phage progeny, as shown schematically in Figure 4. What is seen initially, then, is a variety of different message RNA species. The different message RNA are shown schematically here by a single curve, labeled mRNAi. The mRNA get translated so that after the messages, gene

Table 1. Parameters for bacteriophage T7 and *E. coli.*

Parameter	Value	Reference
Host growth rate (μ)	1.5	
Host volume (V_c)	1.13×10^{-15} L	Donachie and Robinson (1987)
Initial active EcRNAPs per cell (N_p)	992	Bremer and Dennis (1996)
EcRNAP elongation rate (κ_{PE})	50 bps	Bremer and Yuan (1968)
T7RNAP elongation rate (κ_{PE}))	200 bps	Garcia and Molineux (1995)
RNAP spacing requirement (d_E)	176 bp	Dennis and Bremer (1973, 1974)
Active ribosomes per cell (N_r)	20427	Bremer and Dennis (1996)
Ribosome elongation rate (κ_E)	54 bps	Dennis and Bremer (1974)
Ribosome spacing requirement (d_r)	65 bases	Bremer and Dennis (1996)
NTPs per cell (R)	1.4×10^8	Churchward et al. (1982)
Amino acids per cell (P)	1.1×10^9	Donachie (1968)
Host genomes per cell (G_c)	2.3	Cooper and Helmstetter (1968)
Percent of host genome digested by T7	85%	Sadowski and Kerr (1970)
T7 DNA entry	70, 40, and 200 bps	Garcia and Molineux (1995)
Promoter activities		Ikeda (1992); Dayton et al. (1984)
TE efficiency	$\eta_{TE} = 0.99$	Dunn and Studier (1983)
Tϕ efficiency	$\eta_{T\phi} = 0.85$	Macdonald et al. (1993)
T7 mRNA decay rate constant (κ_d^R)	2×10^{-3}/s	Yamada et al. (1974); McCarron and McAllister (1978); Yamada and Nakada (1976)
T7 protein decay rate constant (κ_d^P)	2.8×10^{-5}/s	Lee and Bailey (1984)
EcRNAP and gp0.7 association constant (K_1)	5.5×10^6 M^{-1}	Endy et al. (1997)
EcRNAP and gp2 association constant (K_2)	5.0×10^7 M^{-1}	Endy et al. (1997)
Gp1 and gp3.5 association constant (K_3)	1.087×10^7 M^{-7}	Kumar and Patel (1997)
Replication fork elongation rate (κ_{PD})	370 bps	Rabkin and Richardson (1990)
Procapsid assembly reaction order (D_n)	4.8	Endy (1997)
Procapsid assembly rate constant (κ_a)	4.6×10^{-16} (subunits$^{3.8}$/min	
Procapsid assembly nucleation level (C_n)	6.23×10^{-8} M	Endy (1997)
DNA packaging rate constant (κ_{pack})	467 bps	Son et al. (1993)

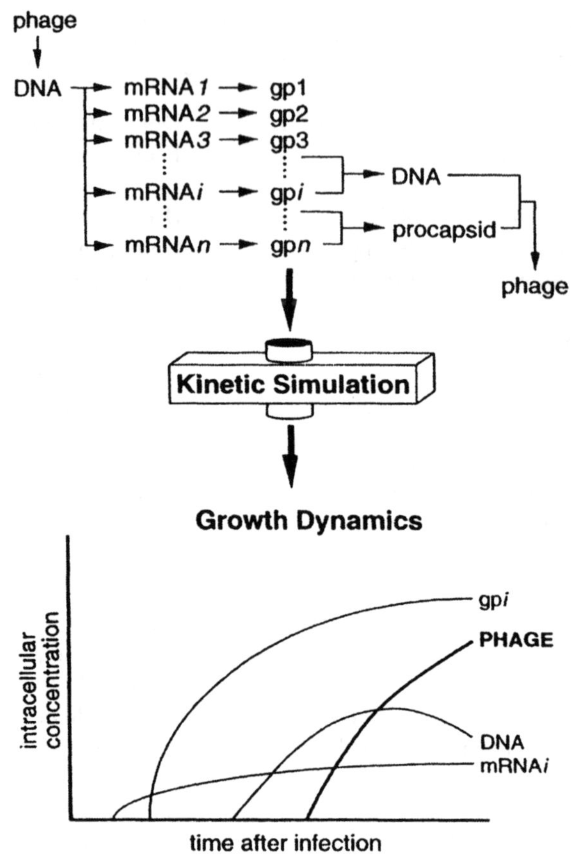

Figure 4. The kinetic simulation takes, as input, the well-established molecular mechanisms and rates of T7 biology. It provides, as output, intracellular concentrations for each T7 component as a function of time post-infection.

products begin to appear (gpi). Some gene products also contribute to the synthesis of T7 DNA. This DNA gets depleted as it is packaged during the process that forms progeny phage.

With phage or other viruses, we may well have some idea of how the information flows. Coupled with quantitative data, this allows us to begin to predict how fast the phage will grow. In essence, we use the simulation to predict growth phenotype from the mechanisms and rates of the constituent reactions. Figure 5 is a comparison of the output of this kind of a simulation with experimental data. On the y-axis are T7 phage particles formed per bacterium, and on the x-axis is the time in minutes post-infection. The experimental points are obtained by infecting a population of cells, breaking them open at different times, and using a biological assay to determine the number of infectious particles formed post-infection. The simulation shown in Figure 5 captures overall the experimentally measured eclipse time and rise rate, but it misses on the burst size. The molecular mechanisms by which the phage actually break out of the host cell are not well understood. When they are better established and incorporated in the simulation we expect predictions for the burst size to improve.

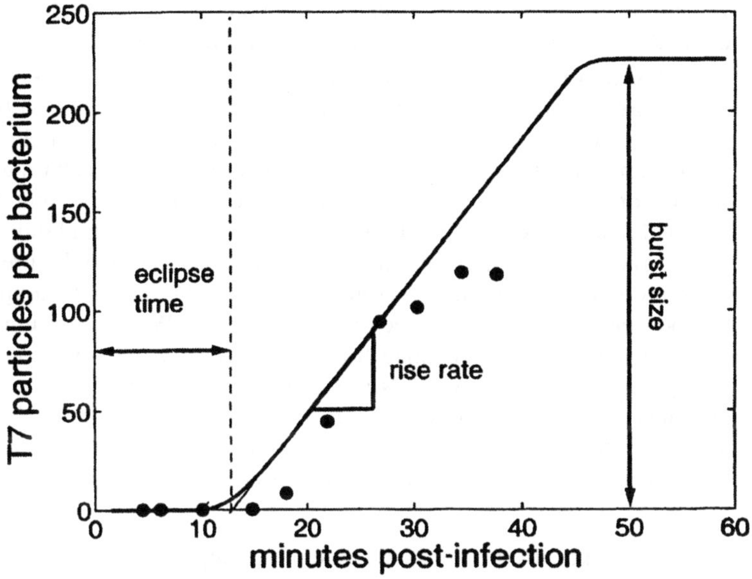

Figure 5. Comparison of simulation and experimental data for T7 growth rate. Host growth rate=1.5 hr^{-1}. The experimental data is shown as data points; the simulated result is shown as a solid line.

Figure 5 shows the results of our initial simulation, implemented by Drew Endy (Endy et al., 1997). Since then, we have used the model to explore a variety of questions. For example, can we identify new kinds of anti-viral strategies? A common strategy is to design the most potent inhibitors of essential virus functions. We have identified strategies that target regulatory loops that might enable us to begin to address the problem of anti-viral resistance to drugs. Such strategies may create a selection against the most common kinds of resistance (Endy and Yin, 2000). We have also shed light on genetic interactions. Population geneticists may be interested in questions of epistasis, which aim to reveal how deleterious mutations interact. For example, do they tend to buffer each other or do they tend to exaggerate their interactions? We have found that the nature of such interactions can depend on the severity of the mutations as well as how one defines fitness (You and Yin, 2002).

4.4. INFERENCE TOOLS

During the development of an organism the various functions encoded in its genome will be activated for transcription and translation at different times. Today this activation is globally assayed using highly parallel gene chip arrays for mRNA and two-dimensional (2-D) gels for proteins. A former graduate student, Lingchong You, asked how the gene-chip information might be related to the 2-D gel information (You and Yin, 2000). If one is rich, then a lot of samples can be taken over time, giving an idea how mRNA and protein levels are changing with time. With our model T7 system, we can readily simulate mRNA levels as well as protein levels as a function of time. We said, let's use our model to "look under the hood." We know what is governing changes in mRNA and protein levels in our model. Can we now take the "raw" mRNA and protein data from our simulation and mine it to reveal underlying mechanisms?

Shown in Figure 6 are some time-series data for T7 mRNA and proteins from our simulation. Levels are in molecules per cell versus time post-infection. Notice that in order to grow, the virus must turn on these various genes. Message RNA and proteins are turned on and there are some twists and turns to the trajectories, but it is hard to infer any sort of mechanisms or functions by looking at these kinds of data. Note that the protein data are not the kind that we would typically get from 2-D gels. The simulation shows the levels of the free proteins, without accounting for the proteins that may be present in the form of a protein-protein complex. In this exercise we are just looking for simple ways to explore the link between the mRNA and the protein trajectory. We have assumed, at least in the simplest case, that we have constant translation rates by the ribosomes, and that translation is only limited by messenger RNA. So we just saturated all the messages with translation machinery and neglected other effects such as distribution of ribosomes, mRNA secondary structure, or codon usage, that might perturb the translation rates. By looking very generically, then, at a protein synthesis rate for a given gene product(i), we have:

$$d(gpi)/dt = k\ [mRNAi] - \text{rate of gpi depletion}$$

where k is the overall translation rate and [mRNAi] is the level of mRNA for gene i.

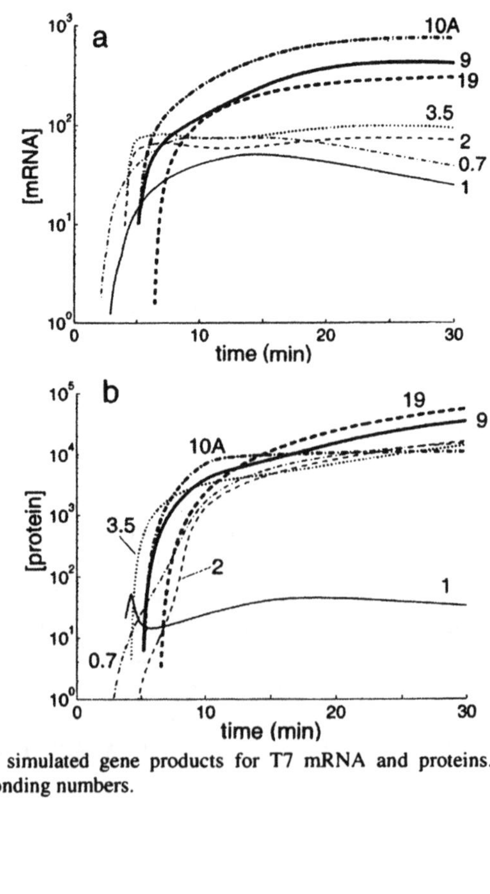

Figure 6. Time-series of simulated gene products for T7 mRNA and proteins. Gene products are labeled according to their corresponding numbers.

Gpi can also be depleted, and this depletion may be due to degradation, formation of protein-protein interactions with other components, or to gpi being shuttled out of the cell. In the very simplest case, in the absence of any depletion term, this expression tells us that protein rates would just depend on translation of messages. By plotting protein rate versus message level, you should just get a straight line having a slope equivalent to the rate of translation. In essence, you can think of the translation machinery of the cell as mathematically integrating the level of messenger RNA over time. The longer we wait, the more protein we will have, and the level over time will also depend on our initial protein level. We can use the raw protein and message data, take a derivative of the protein data as a function of time, and plot it versus mRNA level to see what happens.

We know biology is a little bit more interesting—it's not just all translation. When you plot protein rate versus message level for each gene product, you find some interesting trajectories. These trajectories are demonstrated in Figures 7a and 7b, in which the gene products are labeled by their numbers. Time is implicit here; it starts at zero at the origin and then moves along each trajectory. Gp 10A happens to be the major capsid protein, and its synthesis is seen in the figure. As it deviates from the pure translation line, it reflects the assembly of the procapsids, a process that depletes levels of free protein 10A. If you zoom in on mRNA levels below 100, your find that gp0.7, gp2, and gp3.5 have a number of interesting trajectories that loop back on themselves. We already know these are interesting proteins in T7 biology because we know what we put in the model. In fact, these gene products play a role in feedback loops in the T7 growth, as we see in Figure 8. You can get the main point by tracking gp0.7. The *E. coli* RNAP facilitates transcription of gp0.7, the T7 protein kinase, which then has an inhibitory or negative effect on the *E. coli* RNAP. Negative feedback loops are also active in gp2 and gp3.5. Here we've shown how a detailed model can suggest new ways to do data mining for underlying functions. Note that the mining approach doesn't prove anything about the phage functions, but it does give an idea of which molecular players might be interesting from a regulatory perspective.

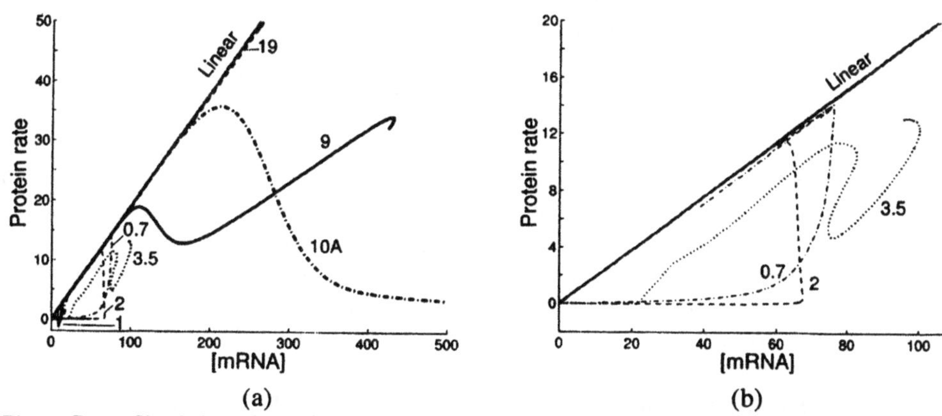

Figure 7. (a) Simulation of protein rate versus mRNA message level for T7, indicating loops defined by regulatory components. Deviation of 10A from pure translation line reflects the depletion of levels of free protein 10A during procapsid assembly. (b) Simulation data for mRNA levels below 100 indicate the role that gene products play in negative feedback loops in T7 growth.

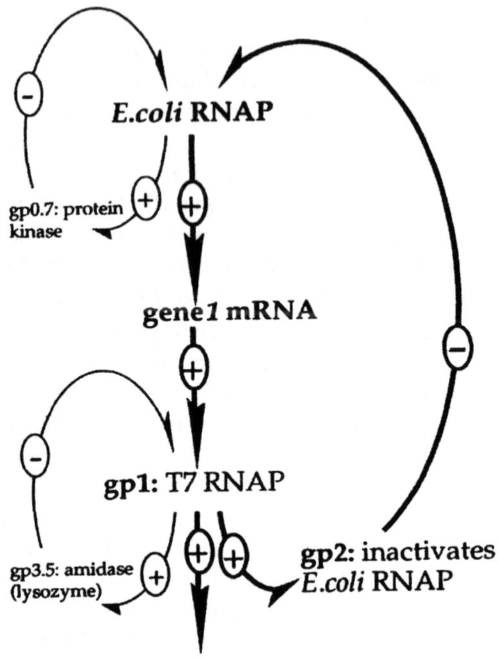

Figure 8. Negative-feedback loops in the early stages of T7 growth.

4.5. NATURE VERSUS NURTURE

A classic question in biology focuses on the roles of nature versus nurture in the development of organisms. Historically, the fundamental studies on viruses have tended to focus on the role of nature, using the characteristics of mutant viruses to deduce virus functions. We have recently employed the simulation to ask about the role of nurture—in essence to ask how the virus growth depends on the resources of its host. The host is being pirated as the virus enters, and its transcription and translation resources are important for the phage growth. So now we split the model into what the phage brings into the cell and what the host cell provides. The host cell may be viewed as a resource reservoir, as shown in Figure 9. This figure shows that the growth rate of the host cell will influence the levels and activities of its various resources, which then serve as inputs into the T7 model, influencing the intracellular rise rate and eclipse times for the phage that we are interested in predicting and understanding. Others have established how the host resources depend on its growth rate (Bremer and Dennis, 1996; Donachie and Robinson, 1987). In essence, we see that by increasing the growth rate of the host, we enrich the resources (nurture) for the phage.

To what extent, then, can the physiological state of the host, characterized by its growth rate, influence the development of the phage? To answer this question we set up flow-through reactors with controllable dilution rates, defined as the flowrate of feed medium to the vessel divided by the liquid volume within the reactor (You et al., 2002). By setting the dilution rate, inoculating the reactor with *E. coli*, and allowing the system

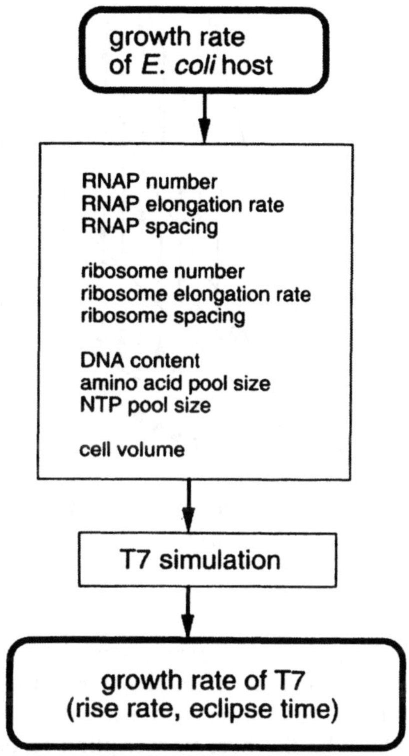

Figure 9. Setting the growth rate of the host cell sets cellular resource levels at values that are incorporated into the simulation and used to predict the growth rate of phage T7.

to reach a steady-state, one can achieve a balanced growth condition where the cells grow at a rate that matches the dilution rate. Hence, by setting different dilution rates, one can "dial up" different cell growth rates and thereby influence the resource environment. Patrick Suthers performed these experiments, using the reactor to prepare hosts at different growth rates, then infected them with phage T7 and determined the phage one-step growth curves.

Figure 10 shows five different growth conditions superimposed. The simulations are shown in the solid lines, and the different experiments, done in triplicate, are shown by the different symbols. As the growth rate of the host cell is increased, the overall trend is that the slope on each of the phage growth curves is also increased. We can extract out this rise rate from those curves to get a rise rate of phage versus growth rate of host, shown in Figure 11. Overall, then, we were able to turn the dial on the host growth rate, change the resources that the host is providing, and see how this is reflected by the processes that the phage is using to grow.

The source of the mismatches in Figure 10 become most apparent when you plot eclipse time versus growth rate, as you can see in Figure 12. Here we see the simulation predicts an earlier appearance of phage than we observed in the experiments. The simplest explanation for this discrepancy is that the initial adsorption of the phage to the host exhibits a longer lag time than implemented in the model. By allowing an adjustment to the lag in the simulation, this discrepancy can be minimized, as indicated

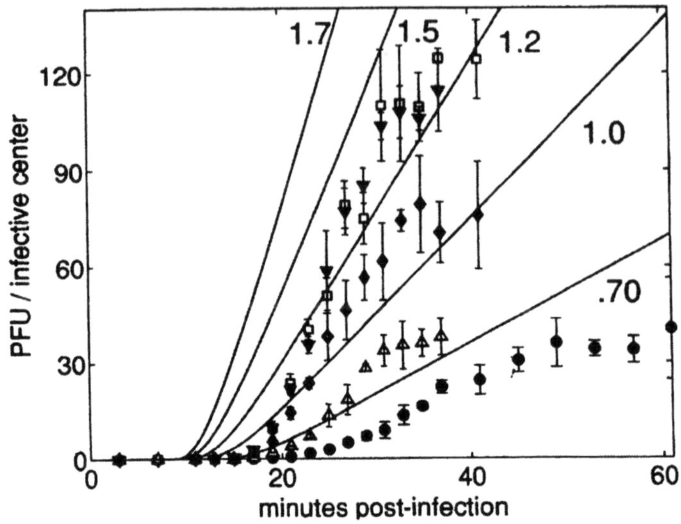

Figure 10. Dependence of phage development on host growth rate. Five different host doubling rates are shown. Solid lines indicate the simulations. Experiments, each done in triplicate, are shown by the different symbols.

by the dashed line. Overall, this set of experiments shows that we have begun to understand how host resource limitations affect phage growth.

Can we take these results a step further and ask: what, if any, specific resources of the host are limiting? Might it be the host transcription resources, precursors to phage DNA, or translation resources? One would ideally like to take an *E. coli* cell and put in 10 more ribosomes or 1,000 more ribosomes without changing anything else in the cell, and then observe the effects on phage growth. This is something that is hard or

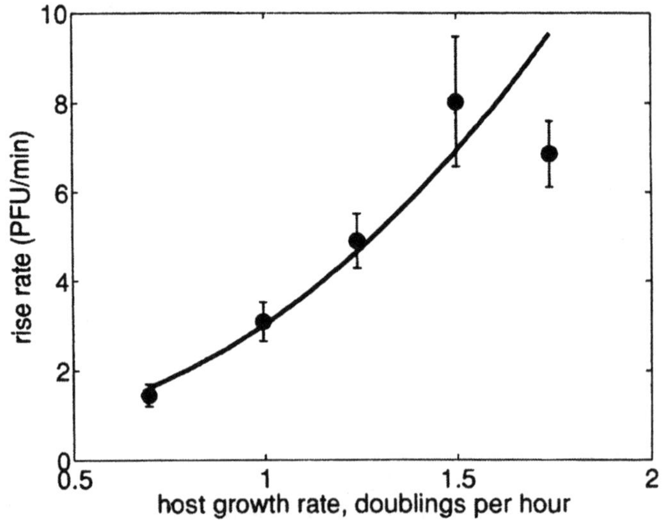

Figure 11. Dependence of rise rate on host growth. The T7 rise rate was extracted from the T7 growth curves in Figure 10, and is shown here in comparison to the host cell growth rate.

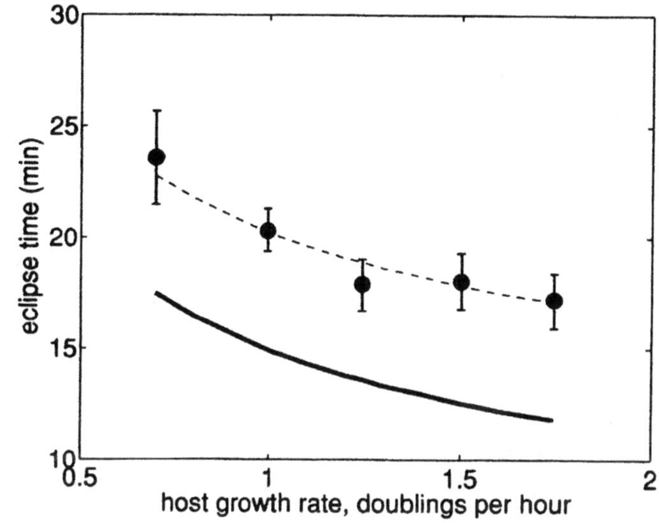

Figure 12. Dependence of eclipse time on host growth.

impossible to do experimentally, but at least the thought experiment can be done using the simulation. One can just say, "I am going to take this representation of my host cell, supply more or less of any one of these resource constituents, and see how it affects the phage growth." Figure 13 shows the results of such an exercise, which was performed by Lingchong You. Here he examined how the rise rate from the one-step growth of the phage depends on levels and activities of various host resources. Quantities on both axes

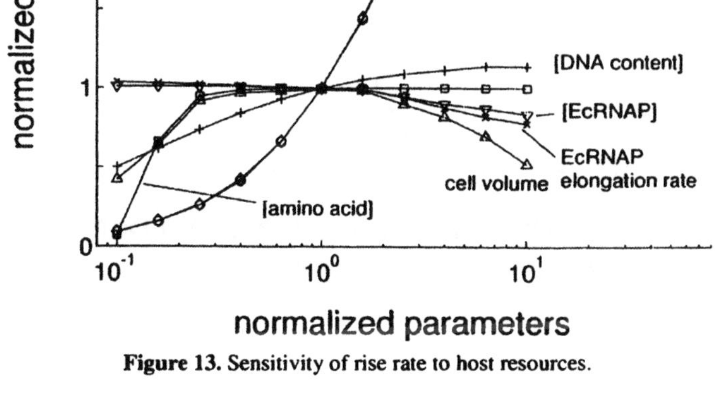

Figure 13. Sensitivity of rise rate to host resources.

are normalized with respect to their wild-type phage or base-case host values. Lingchong found that if he increased the number of ribosomes by 10-fold, then he'd get a much higher rise rate. The phage are growing much better with more ribosomes. It didn't matter much if he increased the DNA content of the cell or the level of the *E. coli* RNA polymerase.

One sees by this kind of plot that the level and processivity of the ribosomes are limiting. Hence, the process of translation is limiting. Will this always be the case? Not necessarily. If the translation resources are increased enough, something else should become limiting. We carried out this exercise and found that if the translation bottleneck is relieved, then growth can become limited by the host transcription machinery, as shown in Figure 14. Further, if we then relieved the transcription bottleneck by, for example, increasing the level of host RNA polymerase, then the bottleneck shifts to the synthesis of DNA for the phage progeny. By asking these questions one begins to create a landscape of potential resource limitations. Now one might say, there's no way the phage will ever encounter a host environment where there are so many more ribosomes or so many more *E. coli* RNA polymerases than the base case. This may be true, but if the phage is a mutant such that its DNA synthesis is slow or deficient in some way, then the DNA-synthesis limiting region in Figure 14 may well expand into the base-case region. In short, for different hosts or different phage, one can imagine these lines and boundaries may move around. Ultimately, the simulation enables us to begin to formulate and better understand how the dynamics of phage development depends not only on the activities of its genome-encoded functions, but also on the available environmental resources.

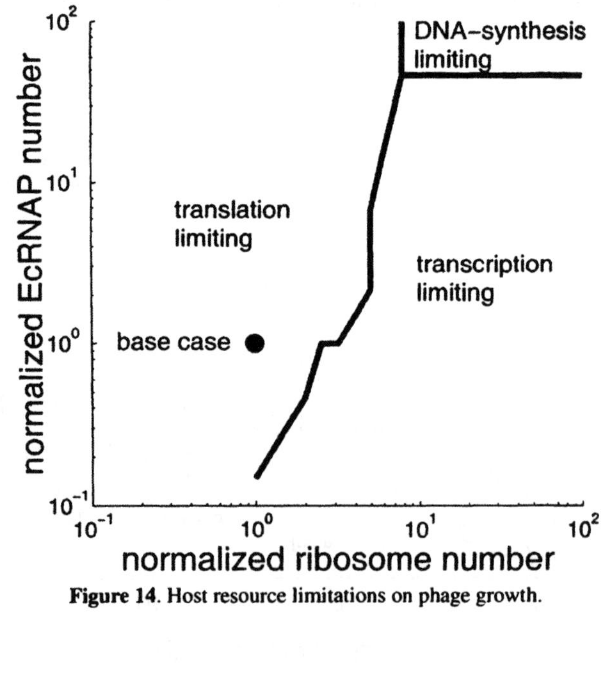

Figure 14. Host resource limitations on phage growth.

4.6. ACKNOWLEDGMENTS

Drew Endy was the first student on this project, which was then taken up by Lingchong You, now a postdoc at Caltech. Drew now has his own lab at MIT. Patrick Suthers, a current graduate student, performed the experimental work. I also thank the National Science Foundation and the Office of Naval Research for financial support for our research.

4.7. REFERENCES

Bremer, H., and Yuan, D., 1968, RNA Chain growth-rate in *Escherichia coli*, *J. Molec. Biol.* **38**:163-180.

Bremer, H., and Dennis, P. P., 1996, Modulation of chemical composition and other parameters of the cell by growth rate, in: *Escherichia coli and Salmonella: Cellular and Molecular Biology*, 2nd Ed. F. C. Neidhardt, R. Curtiss III, J. L. Ingraham, et al., ed., ASM Press, Washington, DC. **II**: pp.1553-1569.

Cooper, S., and Helmstetter, C., 1968, Chromosome replication and the division cycle of Escherichia coli B/r, *J. Mol. Biol.* **31**:519-540.

Dayton, C. J., Prosen, D. E., Parker, K. L., and Cech, C. L., 1984, Kinetic measurements of Escherichia coli RNA polymerase association with bacteriophage T7 early promoters, *J. Biol. Chem.* **259**:1616-1621.

Dennis, P. P., and Bremer, H., 1973, Regulation of ribonucleic acid synthesis in Escherichia coli B/r: An analysis of a shift-up 1. Ribosomal RNA chain growth rates, *J. Molec. Biol.* **75**:145-159.

Dennis, P. P., and Bremer, H., 1974, Macromolecular composition during steady-state growth of Escherichia coli B/r, *J. Bacteriol.* **119**:270-281.

Donachie, W., 1968, Relationships between cell size and time of initiation of DNA replication, *Nature* **219**:1077-1079.

Donachie, W. D., and Robinson, A. C., 1987, Cell division: Parameter values and the process, in: *Escherichia coli and Salmonella typhimurium: Cell. Mol. Biol.* J. L. Ingraham, K. B. Low, B. Magasanik, M. Schaechter, and H. E. Umbarger, eds. ASM Press, Washington, DC. **2**: pp. 1578-1593.

Dunn, J. J., and Studier, F. W., 1983, Complete nucleotide sequence of bacteriophage T7 DNA and the locations of T7 genetic elements, *J. Molec. Biol.* **166**:477-535.

Endy, D., Kong, D., and Yin J., 1997, Intracellular kinetics of a growing virus: A genetically structured simulation for bacteriophage T7, *Biotechnol. and Bioeng.* **55**(2):375-389.

Endy, D., and Yin, J., 2000, Toward antiviral strategies that resist viral escape, *Antimicrob. Agents Chemother.* **44**(4):1097-9.

Garcia, R. L., and Molineux, I. J., 1995, Rate of translocation of bacteriophage T7 DNA across the membranes of Escherichia Coli, *J. Bacteriol.* **177**:4066-4076.

Ikeda, R. A., 1992, The efficiency of promoter clearance distinguishes T7 class II and class III promoters, *J. Biol. Chem.* **267**:11322-11328.

Kumar, A., and Patel, S. S., 1997, Inhibition of T7 RNA polymerase: Transcription initiation and transition from initiations to elongation are inhibited by T7 lysozyme via a ternary complex with RNA polymerase and promoter DNA, *Biochem.* **36**:13954-13962.

Lee, S. B., and Bailey, J. E., 1984, Analysis of growth rate effects on productivity of recombinant Escherichia coli populations using molecular mechanism models, *Biotechnol. Bioeng.* **26**:66-73.

Macdonald, L. E., Zhou, Y., and McAllister, W. T., 1993, Termination and slippage by bacteriophage T7 RNA polymerase, *J. Molec. Biol.* **232**:1030-1047.

McCarron, R. J., and McAllister, W. T., 1978, Effect of ribosomal loading on the structural stability of bacteriophage T7 early messenger RNAs, *Biochem. Biophys. Res. Comm.*

Rabkin, S. D., and Richardson, C. C., 1990, In vivo analysis of the initiation of bacteriophage T7 DNA replication, *Virology* **174**:585-592.

Sadowski, P. D., and Kerr, C., 1970, Degradation of Escherichia coli B deoxyribonucleic acid after infection with deoxyribonucleic acid-defective amber mutants of bacteriophage T7, *J. Virol.* **6**:149-155.

Son, M., Watson, R. H., and Serwer, P., 1993, The direction and rate of bacteriophage T7 DNA packaging in vitro, *Virology* **196**:228-280.

Yamada, Y., Whitaker, P. A., and Nakada, D., 1974, Early to late switch in bacteriophage T7 development: functional decay of T7 early messenger RNA, *J. Mol. Biol.* **89**:293-303.

Yamada, Y., and Nakada, D., 1976, Translation of T7 RNA In Vitro Without Cleavage by RNaseIII, *J. Virol.* **18**:1155-1159.

You, L., and Yin, J., 2000, Patterns of regulation from mRNA and protein time-series, *Metab. Eng.* **2**(3):210-217.

You, L., and Yin, J., 2002, Dependence of epistasis on environment and mutation severity as revealed by in silico mutagenesis of phage T7, *Genetics* **160**:1273-1281.

You, L., Suthers, P., and Yin J., 2002, Effects of *Escherichia coli* physiology on the growth of phage T7 in vivo and in silico, *J. Bacteriol.* **184**(7):1888-1894.

MOLECULAR MACHINES: MULTIPROTEIN COMPLEXES

5. UBIQUITIN-MEDIATED PROTEOLYSIS: AN IDEAL PATHWAY FOR SYSTEMS BIOLOGY ANALYSIS

Martin C. Rechsteiner*

ABSTRACT

Ubiquitin is a small, evolutionarily conserved eukaryotic protein that can be attached to a wide variety of intracellular proteins including itself. Covalent attachment of ubiquitin to other proteins serves various functions, but its major role is to target cellular proteins for destruction. Cellular components that activate, transfer, remove, or simply recognize ubiquitin number in the hundreds, perhaps even in the thousands. In light of this complexity the ubiquitin pathway is ideal for a systems biology approach.

5.1. INTRODUCTION

The 1990s witnessed an increasing appreciation by biochemists, cell biologists, and geneticists of the fact that proteolysis, particularly the ubiquitin (Ub)-mediated pathway, is a major regulatory mechanism. The Ub pathway plays important roles in controlling the cell cycle (Hershko, 1997; Tyers and Jorgensen, 2000), circadian rhythms (Görl et al., 2001), axon guidance (Campbell and Holt, 2001), transcription (Salghetti et al., 2001), and enzyme levels (Kiernan et al., 2001). In view of the growing family of Ub-like proteins (Hochstrasser, 2000; Hay, 2001), it is possible that covalent attachment of Ub, or "Darwin's phosphate" as R. Hampton has called the protein (Pickart, 2001), and its relatives will surpass phosphorylation as a regulatory mechanism. Although Ub serves non-proteolytic roles, such as histone and actin modification (Mimnaugh et al., 2001; Ball et al., 1987), the activation of cell-signaling components (Deng et al., 2000; Wang et al., 2001), endocytosis (Hicke, 2001), and viral budding (Garrus et al., 2001), the protein's major function appears to be targeting proteins for destruction (Hershko and Ciechanover, 1998). To do so, the carboxyl terminus of Ub is activated by an ATP-consuming enzyme (E1) and transferred to one of several small carrier proteins (E2s) in

* Martin C. Rechsteiner, Department of Biochemistry, University of Utah School of Medicine, Salt Lake City, UT 84112.

the form of a reactive thiolester. Ubiquitin is then transferred to lysine amino groups on the proteolytic substrates (S) by one of several large families of Ub ligases or E3s. Chains of Ub are formed, and the conjugated substrate is subsequently hydrolyzed by a large ATP-dependent protease denoted P in Figure 1. The protease, now called the 26S proteasome, was discovered in my laboratory in 1986 (Hough et al., 1986) and purified a year later (Hough et al., 1987).

The Ub system also rivals phosphorylation in the number of components involved in the pathway. It has been estimated that the worm, *C. elegans*, expresses more than 400 protein kinases (Plowman et al., 1999). Similarly, eukaryotic genomes contain information for more than 20 E2s, 100s of E3s, about 100 isopeptidases, and 100s of proteins containing Ub-like domains. In contrast to the wealth of components devoted to marking protein substrates for destruction, there is only one 26S proteasome for actually degrading the ubiquitylated proteins. However, there is complexity here as well because the 26S proteasome is an assemblage of 34 different subunits, and there are four proteasome activators identified to date, with more awaiting discovery. Clearly the Ub system is a rich area for systems biology approaches. It is also too large a field to cover in a single essay or book, for that matter. So I will restrict discussion to the 20S proteasome and the cellular components that bind and activate it.

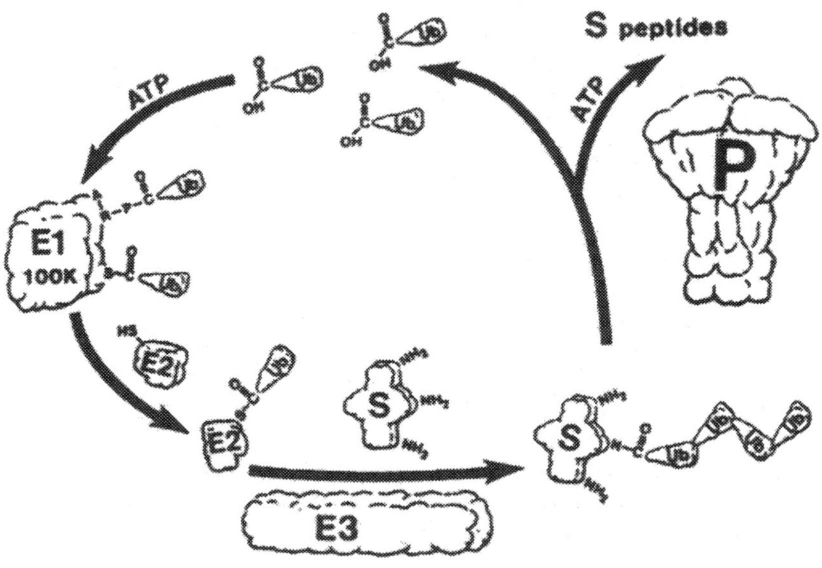

Figure 1. Schematic representation of ubiquitin activation and ATP-dependent proteolysis of conjugated substrates.

5.2. THE 20S PROTEASOME

The 20S proteasome is a major intracellular protease in organisms from all three kingdoms. In eukaryotes the enzyme is present in both nucleus and cytoplasm but not within membrane-bound organelles. Crystal structures have been solved for proteasomes from an archaebacterium and from budding yeast (Baumeister et al., 1998). The 20S proteasome is a cylindrical particle consisting of 28 subunits arranged as four heptameric rings stacked on one another (see Figure 2). The simpler archaebacterial enzyme is composed of just one kind of α subunit and one β subunit, which form the end rings and the two central rings, respectively. Proteolysis is the province of the β subunits, and their active sites face a central chamber within the cylinder. The α subunit rings seal off the central chamber from the external solvent making the 20S proteasome a perfect protease to have among the proteins comprising nucleus and cytoplasm. Unless a native protein is forced into the proteasome interior, it is thought to be impervious to the enzyme. Eukaryotic proteasomes maintain the overall structure of the archaebacterial enzyme, but they exhibit a more complicated subunit composition. There are seven different α subunits and at least seven distinct β subunits. Although current evidence indicates that only three of the seven β subunits are catalytically active, the eukaryotic proteasome cleaves a wider range of peptide bonds. The archaebacterial enzyme has 14 identical copies of a chymotrypsin-like subunit; that is, the enzyme prefers to cleave peptide bonds following hydrophobic amino acids. By contrast, the eukaryotic proteasome contains two

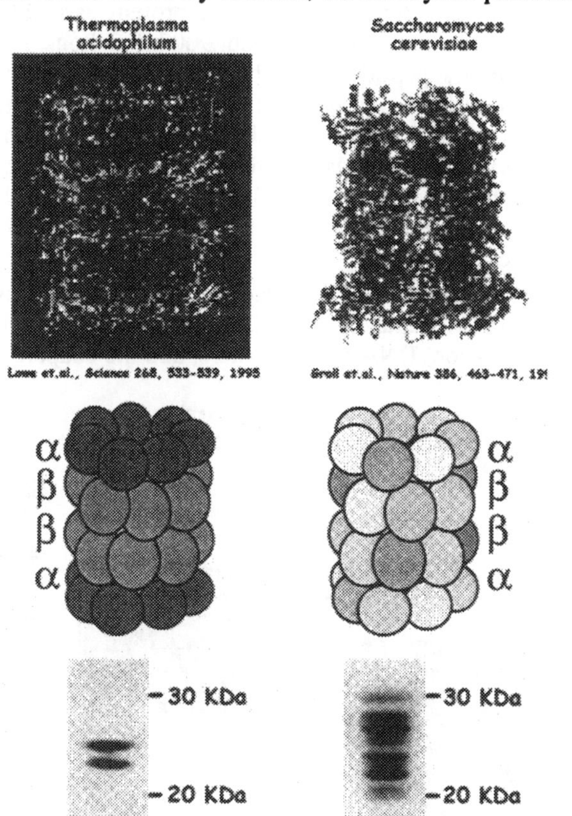

Figure 2. 20S Proteasomes. The top two panels show ribbon diagrams of an archaebacterial proteasome (left) and a eukaryotic proteasome (right) as revealed by x-ray crystallography. The SDS-PAGE patterns of their subunit composition are shown in the bottom panel. Note that the archaebacterial proteasome is assembled from 14 copies of a single α subunit and 14 copies of the same β subunits (left-middle panel). By contrast, the yeast 20S proteasome consists of 2 copies each of 7 different α and 7 different β subunits (right-middle panel).

copies each of trypsin-like, chymotrypsin-like and post-glutamyl-hydrolyzing subunits. Thus it is capable of cleaving almost any peptide bond having difficulty only with Pro-X, Gly-X, and to some extent with Gln-X bonds (Harris et al., 2001).

5.3. THE 26S PROTEASOME

As mentioned, it is evident from x-ray crystallography that the proteasome's internal cavities are virtually inaccessible to intact proteins. To degrade proteins, openings must be generated in the proteasome's outer surface. At present, we know of four protein complexes that bind the proteasome, and three of these are known to promote substrate degradation. The most important of the proteasome-associated components is the 19S regulatory complex (RC), for it combines with the 20S proteasome to form the ATP-dependent enzyme that degrades ubiquitylated and non-ubiquitylated proteins in eukaryotic cells. As the name suggests, this particle is roughly the same size as the 20S proteasome. In fact it is a larger, more complex protein assemblage that contains 18 different subunits with a combined mass of about 900 kilodaltons. Among the 18 subunits are 6 ATPases, a metallo-isopeptidase, a subunit that recognizes polyUb chains, and a number of subunits whose functions are yet to be discovered. As shown in Figure 3, the 18 subunits are arranged in two large sub-complexes called the lid and the base (Glickman et al., 1998). The ATPases and the two largest subunits comprise the base, which sits directly on the end rings of the 20S proteasome; the lid is separated from the base by what appears to be a cavity. Overall, the RC looks much like a Chinese dragonhead (see Figure 3).

The six ATPases in the 19S RC are members of a large family of nucleotidases called the AAA ATPases (Dougan et al., 2002). Other members include NSF, a protein involved in membrane fusion, DNA clamp loaders, and the chaperone p97 to name just a

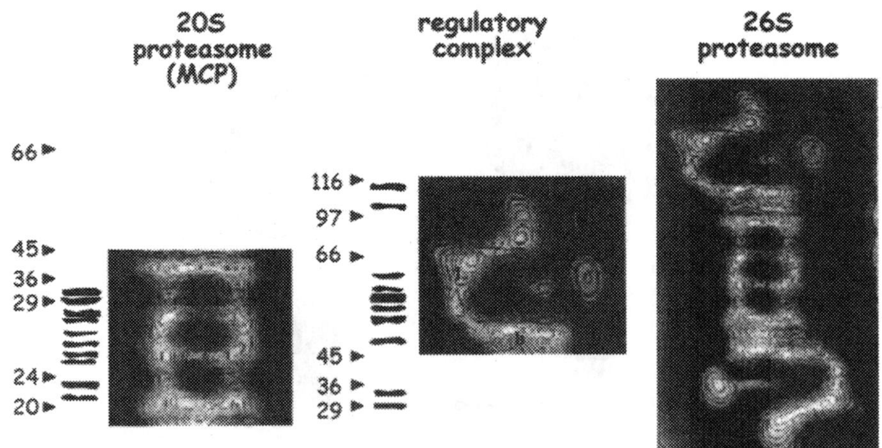

Figure 3. Electron microscopic image of a 26S proteasome. The digitally averaged image depicts two regulatory complexes (RC) bound to opposite ends of the 20S proteasome. The RC can be further divided into two subcomplexes termed the lid (l) and the base (b). The base, which contains the ATPases, S5a, and the two largest subunits, S1 and S2, can activate peptide hydrolysis by the proteasome. However, degradation of polyubiquitylated substrates requires both subcomplexes.

few. All of the AAA ATPases seem to share the common property of altering the association state of proteins or their folded conformation. In the case of the 19S RC ATPases they are thought to act as "unfoldases." Presumably, ubiquitylated proteins are captured by the RC, and the ATPases unfold the bound substrate protein and thread it through the alpha ring of the 20S proteasome for its subsequent degradation in the central proteolytic chamber (see Figure 4). Exactly how the 26S proteasome recognizes polyubiquitylated proteins has not been firmly established. A 50-kilodalton RC protein called S5a has been shown to bind polyUb and polyUb-substrates. However, deletion of the gene encoding this subunit in budding yeast has only minor impact on intracellular proteolysis (van Nocker et al., 1996). Hence, there must be other subunits or other mechanisms by which the 26S enzyme recognizes polyUb-substrates. In this regard, there are an increasing number of reports that the 26S proteasome interacts directly with E3s or E3 binding proteins that are ubiquitylated but not degraded [see Xie and Varshavsky (2002); Alberti et al. (2002)].

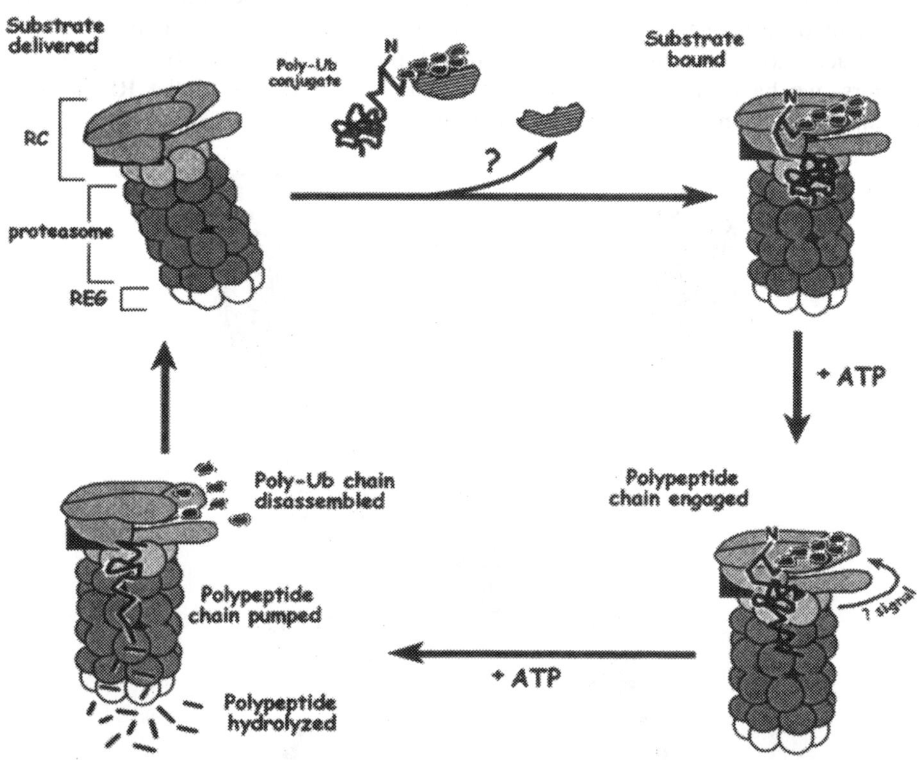

Figure 4. Hypothetical reaction cycle. A polyubiquitylated substrate is delivered to the 26S proteasome possibly by chaperones (step 1). Substrate is bound by polyubiquitin recognition components of the regulatory complex (RC) until the end of the polypeptide chain is engaged by the ATPases (step 2). As the polypeptide chain is unfolded and pumped down the central pore of the proteasome, a signal is conveyed to the S13 isopeptidase to disassemble the polyUb chain (step 3). The unfolded polypeptide is eventually degraded within the inner chamber of the proteasome (step 4).

5.4. PROTEASOME ACTIVATORS

In addition to the RC, there are two protein complexes, REGαβ and REGγ, and a single polypeptide chain, PA200, that bind and activate the proteasome (see Figure 5). There is a fourth protein that binds the proteasome, but it has not yet been demonstrated that this protein, called ecm29 in yeast or golgiPA in mammalian cells, actually enhances peptide bond hydrolysis by the proteasome. Like the RC, proteasome activators bind the ends of the 20S proteasome. And importantly, they can form mixed or hybrid 26S proteasomes in which one end of the 20S proteasome is associated with a 19S RC and the other is bound to a proteasome activator. This latter property raises the possibility that proteasome activators serve to localize the 26S proteasome within eukaryotic cells. In fact, activation per se may be less important than targeting the 26S enzyme to specific locations. On the other hand, it can be argued that failure to quickly release peptide fragments from the central chamber would result in a product-inhibited enzyme. At present we do not know whether activation, targeting, or something else is the major function of proteasome activators. Of course, the answer may vary depending on the specific activator. Whereas we know little of the essential roles of the proteasome activators, we have considerable information on the structures of the REGs and the mechanism by which they activate the proteasome.

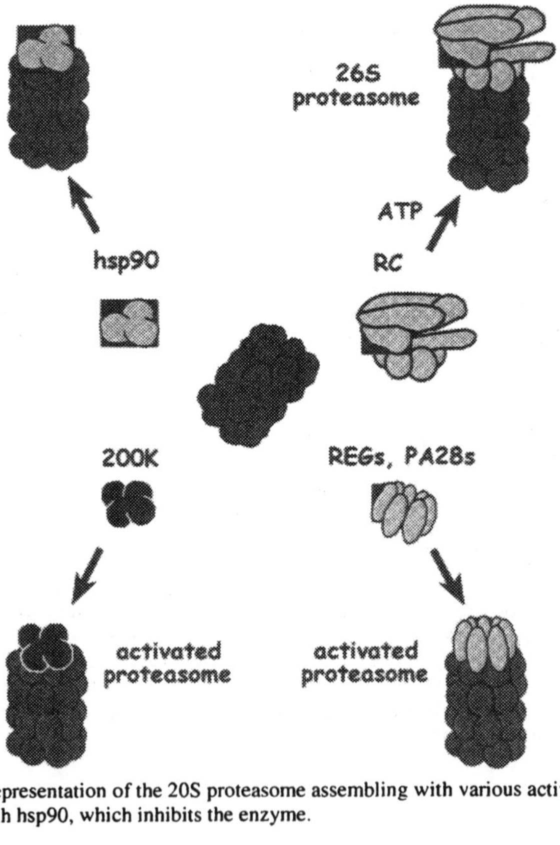

Figure 5. Schematic representation of the 20S proteasome assembling with various activator proteins (RC, REGs, PA200) and with hsp90, which inhibits the enzyme.

5.5. REGS

As depicted in Figure 5, the proteasome is activated by toroidal heptamers called 11S REG, or PA28. There are three distinct REG subunits called αβγ. REGαβ form heteroheptamers found principally in the cytoplasm, whereas REG forms a homoheptamer located in the nucleus. REGαβ are abundantly expressed in immune tissues, while REGγ is highest in the brain. In addition to their different locations, the REGs differ in their activation properties. REGαβ activate all three proteasome active sites; REGγ only activates the trypsin-like subunit. There is reasonably solid evidence that REGαβ play a role in Class 1 antigen presentation, but we have no idea what REGγ does, especially because REGγ knockout mice have virtually no phenotype (Murata et al., 1999). In collaboration with Chris Hill, also at the University of Utah, we solved the structure of the REGα heptamer at 2.8 Å resolution. Seven REGα subunits form a barrel-shaped structure with a central aqueous channel traversing the barrel (Figure 6). The Hill group has also solved the structure of a REG-proteasome complex, which provided important insight into the mechanism of activation (Whitby et al., 2000). Basically, the carboxyl tail on each REG subunit fits into a corresponding cavity on the α ring of the proteasome, and loops on the bottom of the REG subunits displace N-terminal strands on several proteasome α subunits causing them to reorient upward into the aqueous channel of the REG heptamer. These molecular movements open a continuous channel from the exterior solvent to the proteasome central chamber (see Figure 7).

We also have a tentative explanation for the broad activation exhibited by REGαβ and restricted activation of just the trypsin-like subunit seen when REGγ binds. We have obtained mutant REGγs that activate all three active sites. The mutant REGγ heptamers dissociate more readily than wild-type REGγ, leading us to propose that when wild-type REGγ binds to the proteasome, all conformational adjustments occur within the proteasome. We further propose that these conformational adjustments inhibit catalysis at the chymotrypsin-like and post-glutamyl active sites. We speculate that when mutant REGγs bind, they undergo some of the needed conformational adjustment thereby relieving strain on the proteasome active sites.

5.6. PA200

Recently, we reported on a new proteasome activator that we call PA200 based on its molecular weight (Ustrell et al., 2002). Human PA200 is a nuclear protein of 1843 amino acids, and homologs are present in budding yeast, the worm *C. elegans*, and the weed *Arabidopsis*. Apparently *Drosophila* and fission yeast do not have genes encoding PA200. A single chain of PA200 binds the ends of the proteasome and preferentially activates the post-glutamyl active site. A variety of evidence indicates that PA200 is involved in DNA repair. The yeast homolog of PA200 is called Blm3p, so named because mutation of BLM3 confers sensitivity to the DNA damaging agent, bleomycin (Moore, 1991). Yeast cells exposed to another genotoxic chemical, methyl methanesulfonate, up-regulates yPA200 mRNA synthesis about five-fold (Jelinsky et al., 2000). In addition, large-scale proteomic screens revealed that yPA200 associates with the 20S proteasome and the chromatin component Sir4p (Ho et al., 2002). Following

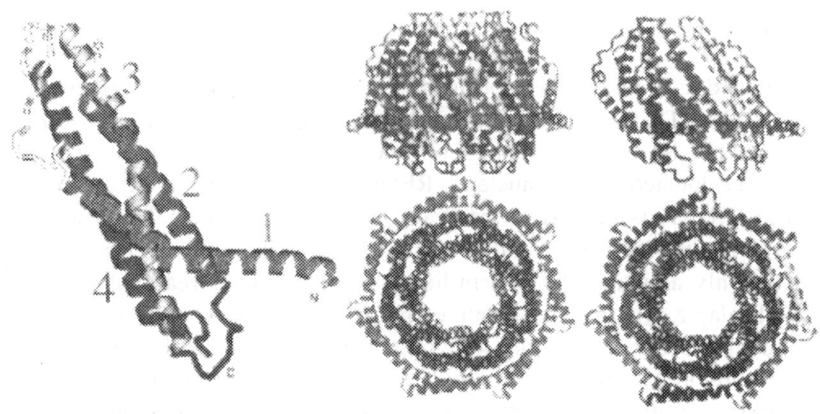

Figure 6. Structure of REGα. At the left is a REGα monomer colored by secondary structural elements. At the right are four views of the REGα heptamer.

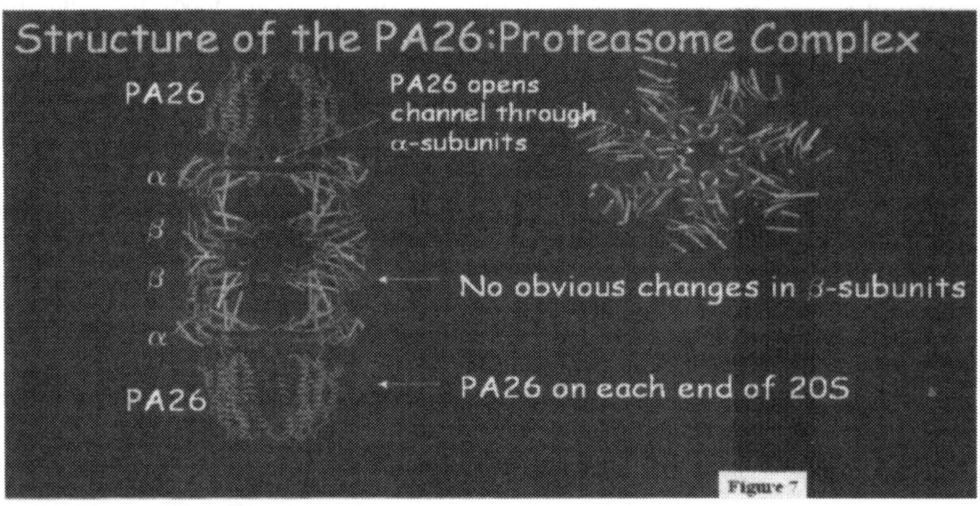

Figure 7. Ribbon representation of the PA26-proteasome complex. Image has been cut away to reveal internal features. Residues of the α-annulus are yellow; ordered N-terminal residues of α-subunits that do not have counterparts in β-subunits are red, and α-subunits are pink.

DNA damage, Sir4p leaves telomeres and relocates to DNA double-strand breaks (DSB) where it binds yKu70, a known component of DSB repair (Martin et al., 1999; Mills et al., 1999). All of these findings strongly implicate PA200 in DNA repair, and this idea is supported by properties of the mammalian PA200. Both PA200 message and protein are abundant in testis, an organ in which DSB occur at high frequency during meiotic recombination. Moreover, HeLa cell PA200 forms intranuclear foci following gamma irradiation as do a number of other repair proteins. So it seems very likely that PA200 is a DNA repair protein, although this cannot be considered a certainty.

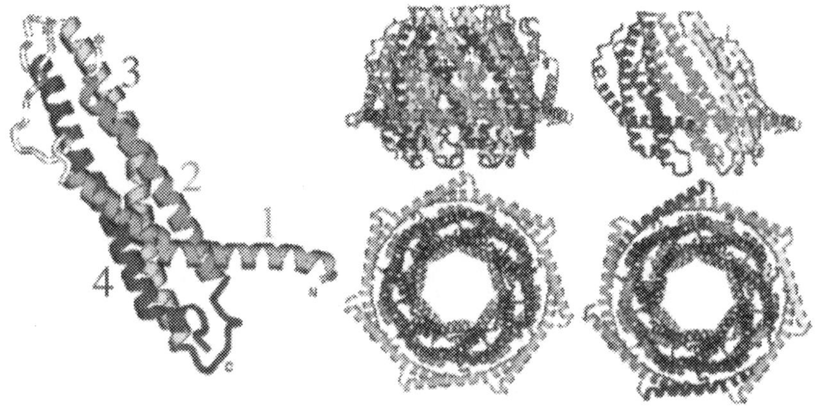

Figure 6. Structure of REGα. At the left is a REGα monomer colored by secondary structural elements. At the right are four views of the REGα heptamer.

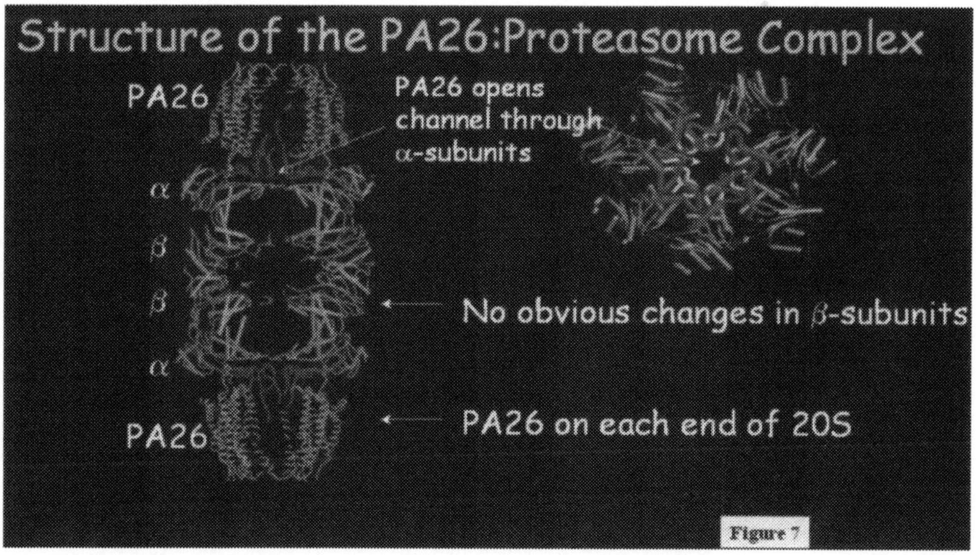

Figure 7. Ribbon representation of the PA26-proteasome complex. Image has been cut away to reveal internal features. Residues of the α-annulus are yellow; ordered N-terminal residues of α-subunits that do not have counterparts in β-subunits are red, and α-subunits are pink.

5.7. ecm29 OR golgiPA

Yet another proteasome-associated protein has recently been discovered. Ecm29 and its homologs in organisms other than yeast are large proteins measuring more than 1800 amino acids. The yeast protein has been identified as a proteasome-associated protein in several proteomic screens (Verma, 2000; Leggett, 2002), and its name derives from the observation that mutation of ecm29 produces defects in the yeast cell wall, hence extracellular mutant (Lussier et al., 1997). Although Leggett et al. have reported that ecm29p serves to stabilize the yeast 26S proteasome (Leggett, 2002), we have found that mammalian ecm29p is found predominately associated with the endoplasmic reticulum golgi intermediate compartment (ERGIC), a location suggesting a role in secretion rather than a general effect on 26S proteasome stability (Gorbea et al., unpublished observation). And location primarily at the ERGIC is more consistent with the yeast ecm phenotype. Whereas it is clearly associated with 20S proteasomes, it is not known if ecm29 actually activates peptide bond hydrolysis by the proteasome. This and a number of other questions concerning this newly discovered protein remain to be answered.

5.8. SUMMARY

I have briefly described the 20S proteasome and the cellular components that bind it. The 19S RC is a special case, as its association with the 20S proteasome produces the 26S proteasome, a cellular particle every bit as essential as the ribosome for cell function. The other proteasome-associated proteins are not essential, and we speculate that they serve to localize the 26S proteasome within eukaryotic cells. Whether this speculation proves to be correct, it is clear that the proteasome interacts with a number of cellular factors and is ripe for detailed systems biology analysis.

5.9. REFERENCES

Alberti, S., Demand, J., Esser, C., Emmerich, N., Schild, H. R., and Hohfeld, J., 2002, Ubiquitylation of BAG-1 suggests a novel regulatory mechanism during the sorting of chaperone substrates to the proteasome, *J. Biol. Chem.* **277**:45920-45927.

Ball, E., Karlik, C. C., Beall, C. J., Saville, D. L., Sparrow, J. C., Bullard, B., and Fyrberg, E. A., 1987, Arthrin, a myofibrillar protein of insect flight muscle, is an actin-ubiquitin conjugate, *Cell* **51**:221-218.

Baumeister, W., Walz, J., Zhül, F., and Seemüller, E., 1998, The proteasome: Paradigm of a self-compartmentalizing protease, *Cell* **92**:367-380.

Campbell, D. S., and Holt, C. E., 2001, Chemotropic responses of retinal growth cones mediated by rapid local protein synthesis and degradation, *Neuron* **32**:1013-1026.

Deng, L., Wang, C., Spencer, E., Yang, L. Y., Braun, A., You, J. X., Slaughter, C., Pickart, C., and Chen, Z. J., 2000, Activation of the IkappaB kinase complex by TRAF6 requires a dimeric ubiquitin-conjugating enzyme complex and a unique polyubiquitin chain, *Cell* **103**:351-361.

Dougan, D. A., Mogk, A., Zeth, K., Turgay, K., and Bukau, B., 2002, AAA+ proteins and substrate recognition, it all depends on their partner in crime, *FEBS Lett* **529**:6-10.

Garrus, J. E., von Schwedler, U. K., Pornillos, O. W., Morham, S. G., Zavitz, K. H., Wang, H. E., Wettstein, D. A., Stray. K. M., Cote, M., Rich, R. L., Myszka, D. G., and Sundquist, W. I., 2001, Tsg101 and the vacuolar protein sorting pathway are essential for HIV-1 budding, *Cell* **107**:55-65.

Glickman, M. H., Rubin, D. M., Coux, O., Wefes, I., Pfeifer, G., Cjeka, Z., Baumeister, W., Fried, V. A., and Finley, D., 1998, A subcomplex of the proteasome regulatory particle required for ubiquitin-conjugate degradation and related to the COP9-signalosome and eIF3, *Cell* **94**:615-623.

Görl, M., Merrow, M., Huttner, B., Johnson, J., Roenneberg, T., and Brunner, M., 2001, A PEST-like element in FREQUENCY determines the length of the circadian period in Neurospora crassa, *EMBO J.* **20**:7074-7084.

Harris, J. L., Alper, P. B., Li, J., Rechsteiner, M., and Backes, B. J., 2001, Substrate specificity of the human proteasome, *Chem. Biol.* **8**:1131-1141.

Hay, R. T., 2001, Protein modification by SUMO, *TIBS* **26**:332-333.

Hershko, A., 1997, Roles of ubiquitin-mediated proteolysis in cell cycle control, *Curr. Opin. Cell Biol.* **9**:788-799.

Hershko, A., and Ciechanover, A, 1998, The ubiquitin system, *Annu. Rev. Biochem.* **67**:425-479.

Hicke, L., 2001, Protein regulation by monoubiquitin, *Nat. Rev. Mol. Cell Biol.* **2**:195-201.

Ho, Y., Gruhler, A., Heilbut, A., Bader, G. D., Moore, L., Adams, S. L., Millar, A., Taylor, P., Bennett, K., Boutilier, K., Yang, L. Y., Wolting, C., Donaldson, I., Schandorff, S., Shewnarane, J., Vo, M., Taggart, J., Goudreault, M., Muskat, B., Alfarano, C., Dewar, D., Lin, Z., Michalickova, K., Willems, A. R., Sassi, H., Nielsen, P. A., Rasmussen, K. J., Andersen, J. R., Johansen, L. E., Hansen, L. H., Jespersen, H., Podtelejnikov, A., Nielsen, E., Crawford, J., Poulsen, V., Sorensen, B. D., Matthiesen, J., Hendrickson, R. C., Gleeson, F., Pawson, T., Moran, M. F., Durocher, D., Mann, M., Hogue, C.W.V., Figeys, D., and Tyers, M., 2002, Systematic identification of protein complexes in Saccharomyces cerevisiae by mass spectrometry, *Nature* **415**:180-183.

Hochstrasser, M., 2000, Evolution and function of ubiquitin-like protein-conjugation systems, *Nat. Cell Biol.* **2**:E153-E157.

Hough, R., Pratt, G., and Rechsteiner, M., 1986, Ubiquitin-lysozyme conjugates. Identification and characterization an ATP-dependent protease from rabbit reticulocyte lysates, *J. Biol. Chem.* **261**:2400-2408.

Hough, R., Pratt, G., and Rechsteiner, M., 1987, Purification of two high molecular weight proteases from rabbit reticulocyte lysate, *J. Biol. Chem.* **262**:8303-8313.

Jelinsky, S. A., Estep, P., Church, G. M., and Samson, L. D., 2000, Regulatory networks revealed by transcriptional profiling of damaged Saccharomyces cerevisiae cells: Rpn4 links base excision repair with proteasomes, *Mol. Cell Biol.* **20**:8157-8167.

Kiernan, R. E., Emiliani, S., Nakayama, E., Castro, A., Labbe, J. C., Lorca, T., Nakayama, K., and Benkirane, A., 2001, Interaction between Cyclin T1 and SCFSKP2 targets CDK9 for ubiquitination and degradation by the proteasome, *Mol. Cell. Biol.* **21**:7956-7970.

Leggett, D. S., Hanna, J., Borodovsky, A., Crosas, B., Schmidt, M., Baker, R. T., Walz, T., Ploegh, H., and Finley, D., 2002, Multiple associated proteins regulate proteasome structure and function, *Mol. Cell* **10**:495-507.

Lussier, M., White, A. M., Sheraton, J., diPaolo, T., Treadwell, J., Southard, S. B., Horenstein, C. I., ChenWeiner, J., Ram, A. F. J., Kapteyn, J. C., Roemer, T. W., Vo, D. H., Bondoc, D. C., Hall, J., Zhong, W. W., Sdicu, A. M., Davies, J., Klis, F. M., Robbins, P. W., and Bussey, H., 1997, Large scale identification of genes involved in cell surface biosynthesis and architecture in Saccharomyces cerevisiae, *Genetics* **147**:435-450.

Martin, S. G., Laroche, T., Suka, N., Grunstein, M., and Gasser, S. M., 1999, Relocalization of telomeric Ku and SIR proteins in response to DNA strand breaks in yeast, *Cell* **97**:621-633.

Mills, K. D., Sinclair, D. A., and Guarente, L., 1999, MEC1-dependent redistribution of the Sir3 silencing protein from telomeres to DNA double-strand breaks, *Cell* **97**:609-620.

Mimnaugh, E. G., Kayastha, G., McGovern, N. B., Hwang, S. G., Marcu, M. G., Trepel, J., Cai, S. Y., Marchesi, V. T., and Neckers, L., 2001, Caspase-dependent deubiquitination of monoubiquitinated nucleosomal histone H2A induced by diverse apoptogenic stimuli, *Cell Death Differ.* **8**:1182-1196.

Moore, C. W., 1991, Further characterizations of bleomycin-sensitive (blm) mutants of Saccharomyces cerevisiae with implications for a radiomimetic model, *J. Bacteriol.* **173**:3605-3608.

Murata, S., Kawahara, H., Tohma, S., Yamamoto, K., Kasahara, M., Nabeshima, Y., Tanaka, K., and Chiba, T., 1999, Growth retardation in mice lacking the proteasome activator PA28gamma, *J. Biol. Chem.* **274**:38211-38211 5.

Pickart, C. M., 2001, Ubiquitin enters the new millennium, *Mol. Cell* **8**:499-504.

Plowman, G. D., Sudarsanam, S., Bingham, J., Whyte, D., and Hunter, T., 1999, The protein kinases of Caenorhabditis elegans: a model for signal transduction in multicellular organisms, *Proc. Natl. Acad. Sci. USA* **96**:13603-13610.

Salghetti, S. E., Caudy, A. A., Chenoweth, J. G., and Tansey, W. P., 2001, Regulation of transcriptional activation domain function by ubiquitin, *Science* **293**:1651-1653.

Tyers, M., and Jorgensen, P., 2000, Proteolysis and the cell cycle: With this RING I do thee destroy, *Curr. Opin. Genet. Dev.* **10**:54-64.

Ustrell, V., Hoffman, L., Pratt, G., and Rechsteiner, M., 2002, PA200, a nuclear proteasome activator involved in DNA repair, *Embo. J.* **21**:3516-3525.

van Nocker, S., Sadis, S., Rubin, D. M., Glickman, M., Fu, H. Y., Coux, O., Wefes, I., Finley, D., and Vierstra, R. D., 1996, The Multiubiquitin-chain-binding protein Mcb1 is a component of the 26S proteasome in Saccharomyces cerevisiae and plays a nonessentail, substrate-specific role in protein turnover, *Mol. Cell Biol.* **16**:6020-6028.

Verma, R., Chen, S., Feldman, R., Schieltz, D., Yates, J., Dohmen, T., and Deshaies, R. J., 2000, Proteasomal proteomics: identification of nucleotide-sensitive proteasome-interacting proteins by mass spectrometric analysis of affinity-purified proteasomes, *Mol. Biol. Cell* **11**:3425-3439.

Wang, C., Deng, L., Hong, M., Akkaraju, G. R., Inoue, J., and Chen, Z. J. J., 2001, TAK1 is a ubiquitin-dependent kinase of MKK and IKK, *Nature* **412**:346-351.

Whitby, F. G., Masters, E. I., Kramer, L., Knowlton, J. R., Yao, Y., Wang, C. C., and Hill, C. P., 2000, Structural basis for the activation of 20 S proteasomes by 11 S regulators, *Nature* **408**:115-120.

Xie, Y., and Varshavsky, A., 2002, UFD4 lacking the proteasome-binding region catalyses ubiquitination but is impaired in proteolysis, *Nat. Cell. Biol.* **4**:1003-1007.

6. CONFORMATIONAL SWITCHING IN MUSCLE

Piotr Fajer*

ABSTRACT

Muscles can be studied as complex systems of many interacting proteins and investigated at many different levels of organization. This talk will describe how we modeled the mechanism of Ca activation, the structure of the muscle proteins, and protein complexes (including actin monomers, tropomysin and troponin complexes, and myosin) to examine two different scientific problems: the mechano-chemical energy transduction mechanism, and the control system of that mechanism. The methods we used—saturation transfer electron paramagnetic resonance, phosphorescence anisotropy, and fluorescence resonance energy transfer—reveal two specific structures: a hinge between the motor and regulatory domains, and a stiff regulatory domain. This indicates that the structure of the myosin head is capable of generating translating conformational changes within the motor domain to the swing of the regulatory domain, and that the regulatory domain is rigid enough to act as a lever arm.

6.1. INTRODUCTION

I want to talk today about muscle and conformational switching. In the 1960s, Hugh Huxley proposed a model for force generation in muscle (Huxley, 1969). Since then, my lab and many others have been working to validate it. Sadly, we still don't have the answer. When I started in the field, I was interested not in muscle biology per se, but in the fact that it is a complex system of many interacting proteins that can be investigated at many different levels of organization. For instance, any particular protein is part of a complex; for example, the protein may be a monomer of actin, one of the subunits of troponin, or one of the light chains of myosin. These proteins can be studied separately or in complexes with other subunits. One can even reconstitute a working muscle fiber with structure, as shown in Figure 1. Muscle fibers that we have reconstituted with proteins of

* Piotr Fajer, Florida State University, Tallahassee, FL 32310.

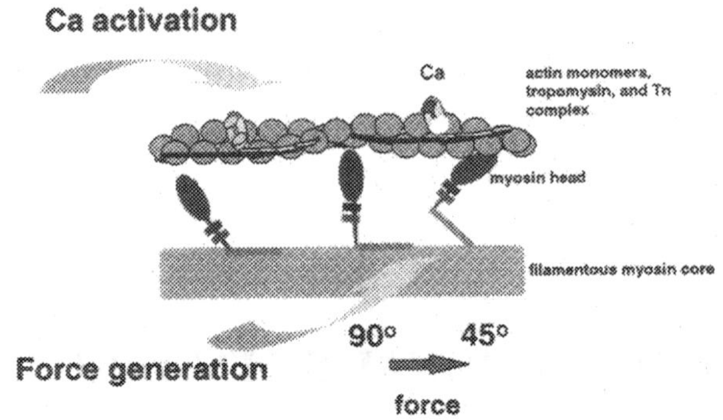

Figure 1. Model of muscle fiber structure indicating various protein-making thin and thick filaments. Thick filaments made of myosin form the filamentous core (*gray*). The globular myosin heads protruding from the filament surface are indicated in magenta. Above the myosin heads are the thin filaments made of actin monomers (*cyan*), tropomyosin (*blue*) and troponin (Tn) complex: TnC (*yellow*), TnI (*pink*), and TnT (*light blue*). Activation of the muscle is initiated by the binding of calcium to TnC. Force generation is thought to be due to rotation of the globular head (Huxley, 1969).

interest are a working unit that can generate force. Permeabilization of the muscle cell's membrane allows us to easily. change its chemical composition; for example, the concentration of ATP or of Ca^{2+}.

An equally interesting aspect of muscle is that, like any motor—biological or mechanical—it must have both an engine and an ignition system. Thus, in one system we have two different scientific problems: the first is mechano-chemical energy transduction, and the second is a control system. The energy transduction problem boils down to one question: How is the force generated? It's probably generated by a conformational change of the myosin head, such as the head rotation seen in Figure 1. Rotation of the head while it is attached to the actin filament would generate a strain between the thick and thin filaments—strain that is the force generated by muscle. Assuming that we know how the engine works, then how do we tell it to start or stop? We know that muscle activation is initiated by the binding of calcium to troponin C (TnC), a part of the thin filament (Figure 1). The conformational changes induced in TnC by calcium binding must propagate through the other troponin subunits, tropomyosin and actin, all the way to the active site of the myosin head where it accelerates the hydrolysis of ATP. Again, it's a complex pathway of interactions involving many different proteins.

The nice thing about muscle research is that, from early on, the effort has been multidisciplinary and driven a lot of instrumental development. If you like gadgets, then this is your playground.

6.2. SPECTROSCOPIC TECHNIQUES IN MUSCLE CONTRACTION

Figure 2 shows some of the spectroscopic techniques that are used to analyze what conformational changes take place when muscle contracts. For example, these techniques are used to determine the distances between protein domains or protein subunits. We use fluorescent resonance energy transfer (FRET) and dipolar electron paramagnetic resonance (EPR). To measure protein dynamics we use fluorescence or phosphorescence anisotropy and saturation transfer electron paramagnetic resonance (ST-EPR). We can

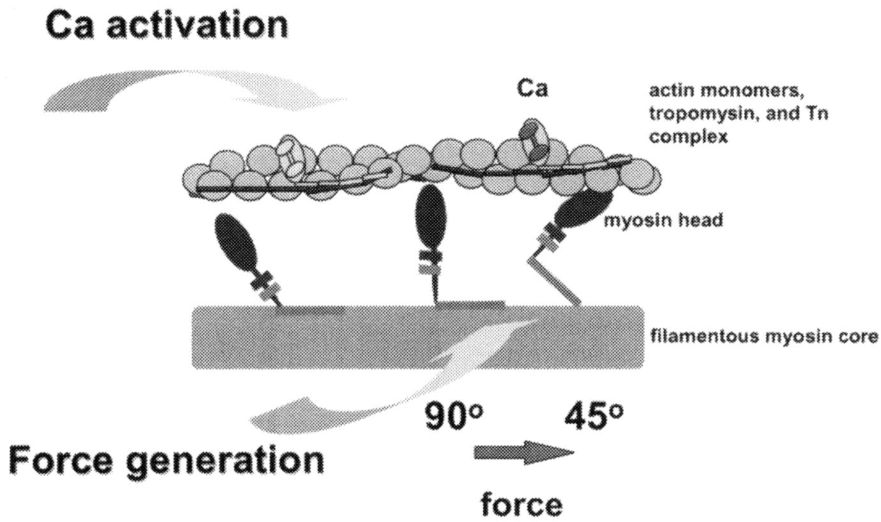

Figure 1. Model of muscle fiber structure indicating various protein-making thin and thick filaments. Thick filaments made of myosin form the filamentous core (*gray*). The globular myosin heads protruding from the filament surface are indicated in magenta. Above the myosin heads are the thin filaments made of actin monomers (*cyan*), tropomyosin (*blue*) and troponin (Tn) complex: TnC (*yellow*), TnI (*pink*), and TnT (*light blue*). Activation of the muscle is initiated by the binding of calcium to TnC. Force generation is thought to be due to rotation of the globular head (Huxley, 1969).

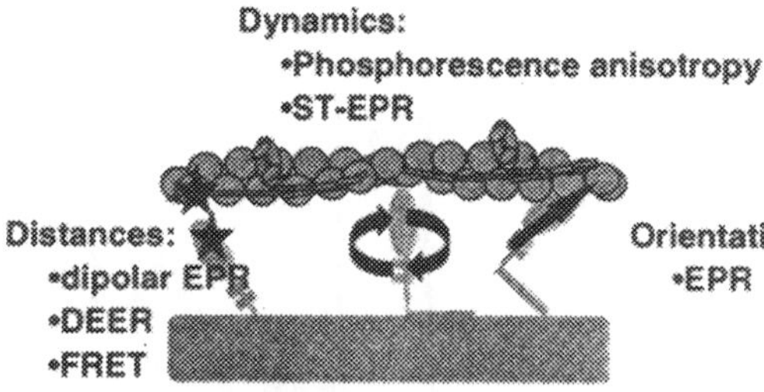

Figure 2. Multiple spectroscopic techniques applied in the investigation of protein distances (*stars*), dynamics (*rotating arrows*), and orientation (*straight arrow*).

even measure the orientation of the proteins within muscle fiber because muscle is an oriented system. EPR is exquisitely sensitive to the orientation between the external magnetic field and a spin probe attached to a selected protein.

Figure 3 shows an EPR spectrum of labeled muscle. From the splitting of the signal you can measure an average angle of the label with respect to fiber axis. Furthermore, from the width of this signal, you can determine the orientational disorder of your labeled proteins. Direct measurement of molecular disorder is of interest because order-disorder transitions have been increasingly implicated in the mechanism of molecular action. EPR measures molecular disorder directly; the broader the resonance, the larger the disorder of the spins, the wider the distribution about the average angle. As an aside, EPR is one of the very few techniques, if not the only one, that displays disorder directly, rather than indirectly by the absence of the expected signals. For example, in x-ray crystallography disorder is displayed indirectly by absence of electron density, in NMR by absence of NOE resonances. EPR sums all the signals, good, bad, and ugly, rather than averaging them, so that all the orientations are represented in the EPR spectrum.

Dipolar EPR can measure distances in a similar way to FRET. Dipolar EPR is based on spin-spin interaction analogous to the donor-acceptor interaction in FRET. Two labels talk to each other if the distance between them is short enough. The interaction expresses itself in the broadening of an otherwise sharp signal. Figure 4 demonstrates this

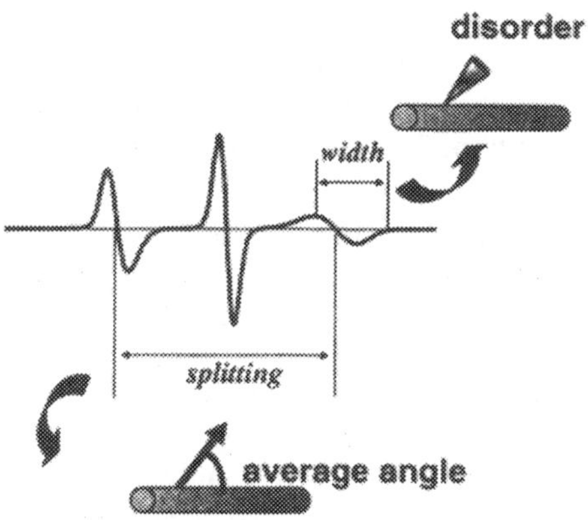

Figure 3. The EPR spectrum is sensitive to protein disorder and orientation. Orientation determines the splitting; disorder defines width of the resonance.

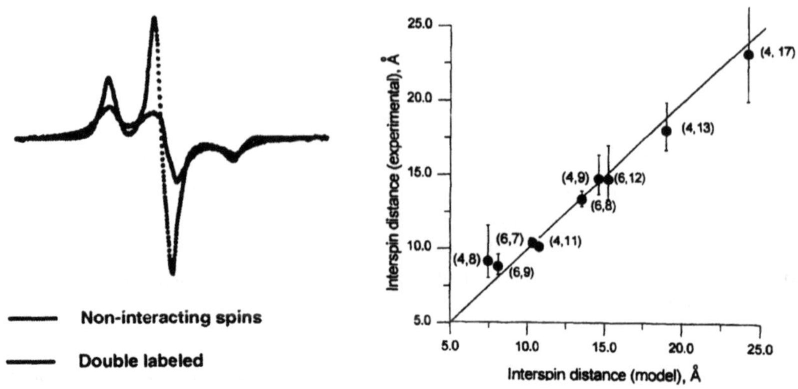

Figure 4. Dipolar EPR sensitivity to distances between two interacting spin labels. The spectrum is from a protein containing a single spin label is indicated by the solid line. The same spectrum broadened by the presence of a second label is indicated by the dotted line. The graph on the right compares known distances for doubly labeled synthetic peptides with the distances obtained from dipolar EPR (Rabenstein and Shin, 1995).

interaction: the spectrum from a single spin label is sharp; if two labels are in proximity to each other, the spectrum is broadened. The broadening of that spectrum is related to the spin-spin distance.

The graph on the right side of Figure 4 verifies that the distances obtained from the broadening of the spectra correspond to the interspin distance. A few years ago, Rabenstein and Shin (1995) synthesized peptides with two cysteines separated by a varying number of residues. They attached spin labels to the cysteines, and calculated the interspin distance from the broadening. The comparison of the observed distance and separation distance along the peptide was excellent for distances between 8 and 25Ångstroms. This usable range can be expanded to 80 Ångstroms using pulsed EPR, a technique developed by Milov et al. (2000) and by Jeschke (2002). The method—a double-resonance-pulsed EPR—is somewhat similar to the two-channel NMR methods but technically more difficult (because the relaxation times are faster than in NMR, the timing must be faster). Figure 5 shows the formation of an echo following two pulses. The third pulse is applied at a slightly different frequency, which causes it to excite a different spin label. The position of this pulse is swept between the two previous pulses. This leads to the change of the intensity of the echo modulation that when Fourier transformed yields a Pake pattern. The splitting of the Pake pattern is directly related to the inter-spin distance.

In all of these spectroscopic techniques—fluorescence (FRET), phosphorescence, or EPR—we need to introduce labels into defined positions on the protein surface. Most of the labels modify cysteines residues. This means that we either have to rely on the availability of naturally occurring cysteines or we have to engineer them into the protein using mutagenesis. The latter becomes easier every year; for example, in one project, 20 residues of troponin I (TnI) were, in turn, changed into cysteines. Figure 6 shows the variety of labeling sites on the myosin head used in this study; some of them are naturally occurring, some are engineered. There is also a substantial variety in the linkages between the protein and the nitroxide moiety that is a stable radical giving rise to an EPR signal. As seen in Figure 6, we have iodoacetamide, maleimide, indane dione, and thiosulfonate linkages, to name a few. By varying the label on any given site, we can vary

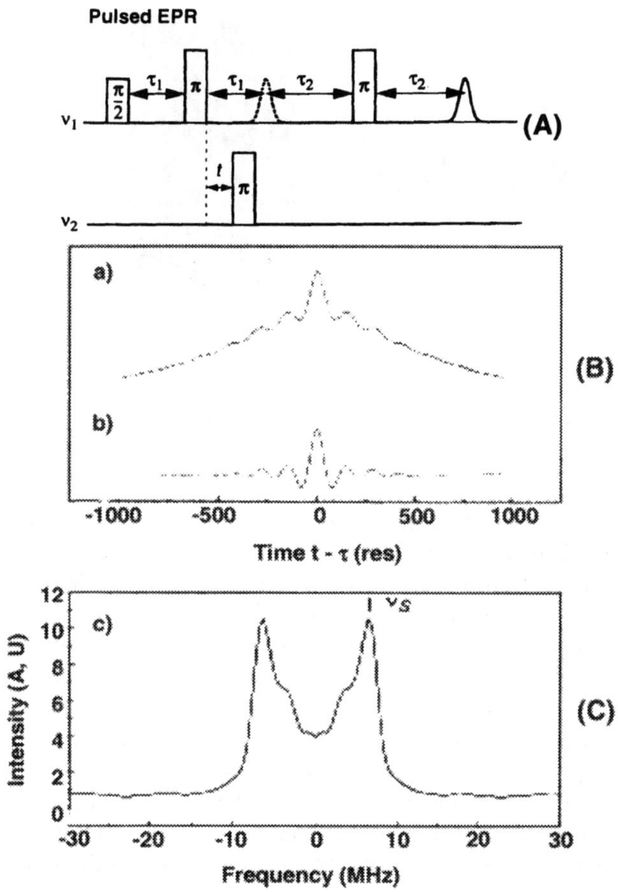

Figure 5. Double Electron Electron Resonance (DEER). (A) the pulse sequence of the microwave at observing frequency (v_1) and saturating pulse at (v_2) that create an echo; (B) the modulation of the echo intensity before and after subtraction of relaxation; (C) pake pattern created by Fourier transformation of the echo modulation. The splitting of the Pake pattern is determined directly by the spin-spin distance (Jeschke, 2002).

its orientation or its mobility with respect to the protein. Large linkages or linkages containing double bonds will generally be more rigid than the labels attached by long linkers with 4-5 single bonds.

In some cases, a stiff linkage facilitates a stereo-specific attachment that allows measurement of the orientation of the labeled domain. The process required to determine the domain orientation from the EPR spectrum of the spin labels, however, is not trivial. The EPR spectrum gives us the orientation of the radical with respect to the magnetic field. For ordered proteins, planar lipid bilayers, or the cylindrically ordered muscle, the orientation with respect to the magnetic field translates into orientation with respect to the symmetry axis of the sample. However, what we really want to know is the orientation of a protein domain with respect to the macromolecular assembly.

So what can we do? If you know the orientation or motion of the spin label with respect to the protein, then you can use it to translate the label orientation with respect to the macromolecular symmetry axis into protein orientation with respect to the assembly.

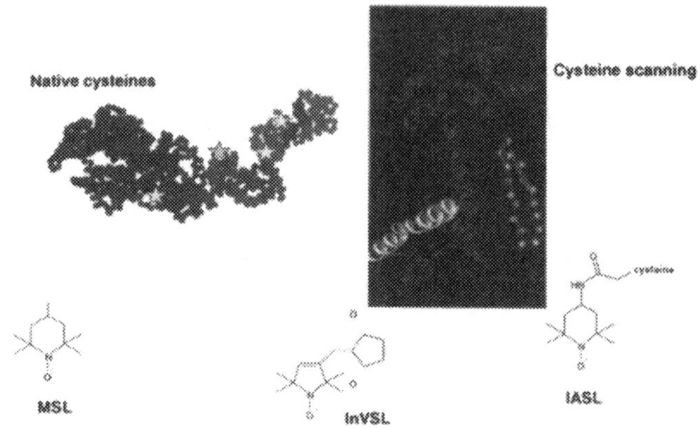

Figure 6. Various spin labels and labeling sites on the myosin head *(left)* and sites engineered by cysteine scanning of TnI *(right)*.

The answer, therefore, is to find the orientation of the label with respect to the protein. We have accomplished this using crystal structures of the labeled environment and an extensive conformer search. Figure 7 is a standard Metropolis Monte Carlo simulation. We generate a random conformer of the label, single bonds are all set to a random value, we minimize the structure, calculate the energy, compare to a previous lowest energy structure, and using the Metropolis criterion, either: keep it or ignore it and search for the next conformer.

This is of course an *in silico* experiment, and though I like computers, even I don't quite trust them. What we need is a validation system to test the simulation. Then, we might be able to actually convince ourselves that the *in silico* fantasies work. Our validation system is a muscle fiber with a myosin head attached to an actin filament in the absence of nucleotides. Electron microscopy image reconstructions combined with x-ray

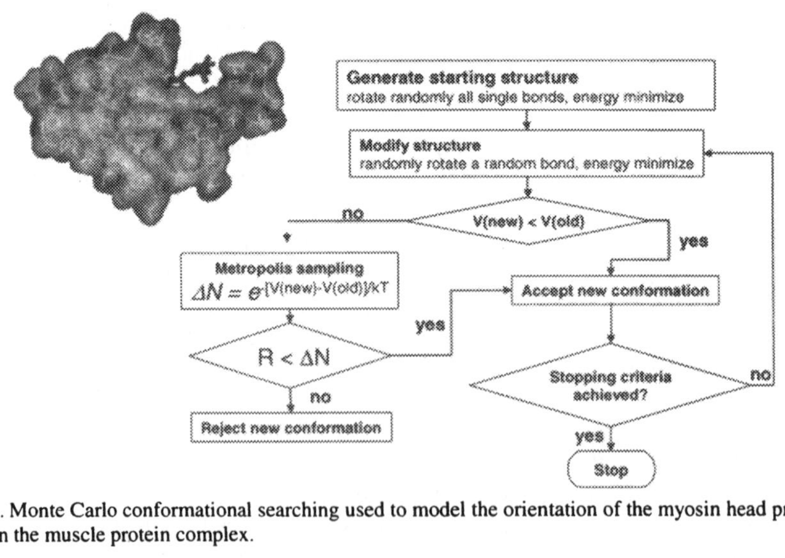

Figure 7. Monte Carlo conformational searching used to model the orientation of the myosin head protein domain in the muscle protein complex.

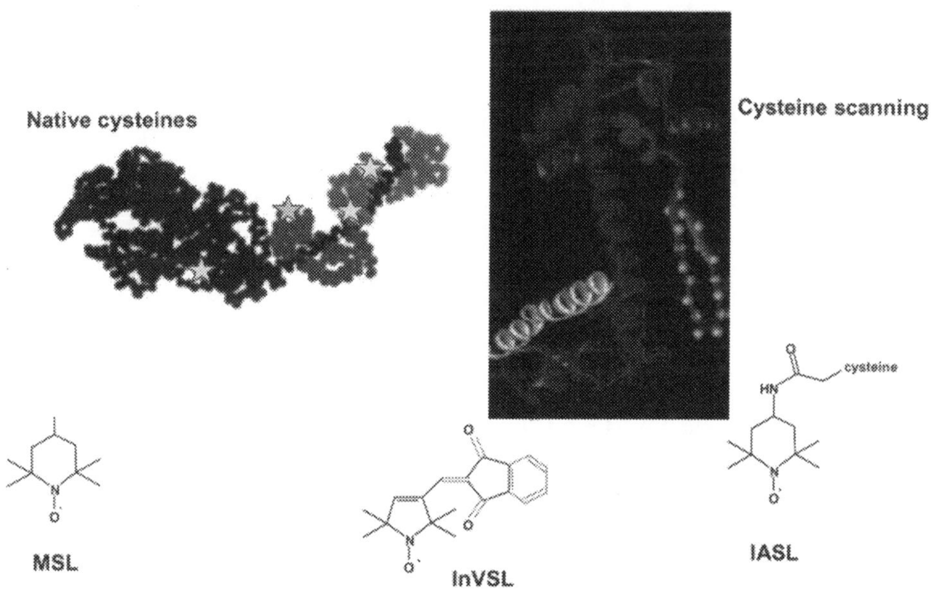

Figure 6. Various spin labels and labeling sites on the myosin head *(left)* and sites engineered by cysteine scanning of TnI (*right*).

atomic-level structures have determined with high accuracy the orientation of the large, asymmetric motor domain of myosin head (Figure 8). This is the "poor man's" crystal. Myosin heads are oriented cylindrically within a crystallographic unit cell—muscle fiber, and we know that orientation. We can, therefore, predict what the EPR spectrum should look like based on the predicted orientation of the label with respect to the protein and the known orientation of the protein with respect to fiber axis. More important, we can observe such a spectrum and correlate the predicted and observed orientations. Figure 8 illustrates this comparison. On one axis, we have the predicted orientation of six different labels; on the other axis we have the same orientation calculated from the observed spectra. The agreement is very good (Sale et al., 2002).

In addition to predicting the label orientation, we are trying to predict its mobility on the protein surface. Spectroscopic measurements of molecular dynamics reflect the dynamics of the probe rather than the molecule. If the probe is rigidly attached to a protein, then probe movement reflects protein movement. If the probe moves with respect to the protein, then the observed mobility consists of a combination of (1) the protein domain movement that is of biological interest, and (2) the probe floppiness that is an unwanted byproduct of the labeling approach. If we could deconvolve label floppiness from the observed spectra, we would be left with the protein mobility. Our approach to determine the label mobility is to use molecular dynamics simulations of the label relative to the protein surface. Trajectories are then converted into amplitudes and rates of motion that are used as input into spectral simulations. Again, validation is everything. Figure 9 shows an EPR spectrum of label mobility with respect to the protein. The solid line is the experimental spectrum, and the dotted line is a spectrum simulated from molecular dynamics calculations. That agreement is quite promising. My goal here was to

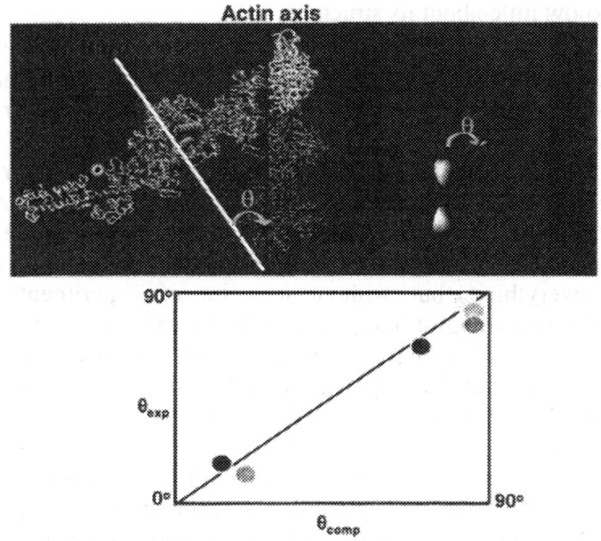

Figure 8. Validation of the Monte Carlo-determined label orientation (*abscissa*) and the orientation inferred from the EPR spectra (*ordinate*). The top graphic shows the model system used in the validation. The orientation of the myosin head is known from electron microscopy that is translated into the orientation of the probe with respect to the magnetic field.

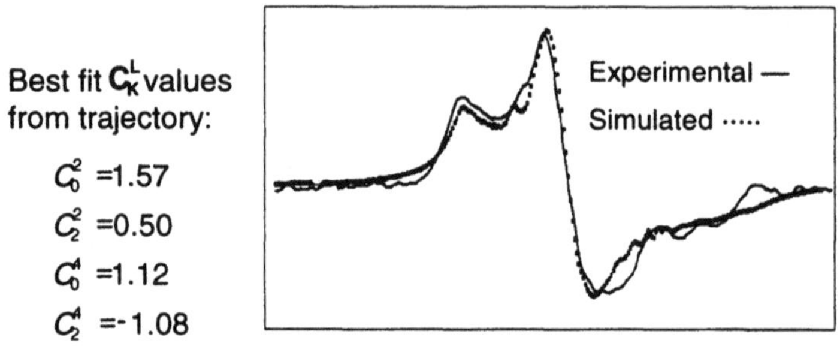

Figure 9. Prediction of the spin mobility by molecular dynamics simulations. The dotted line indicates the EPR spectrum simulated using amplitude of motion from MD trajectories. The solid line indicates the experimental spectrum.

provide a flavor of how to interpret the spectrum and convince you that spectroscopy, under some circumstances, could be useful in determining the orientation or the dynamics of the protein domains.

6.3. STRUCTURE OF THE INHIBITORY REGION OF TROPONIN

So what are technique and technique development good for? After all, even the most elegant techniques head for a landfill if they can't find exciting applications. In our case, an exciting application is the determination of the structure of the inhibitory region of troponin. This is the region that is directly involved in the inhibition of force generation. A synthetic peptide with a similar sequence is capable of force inhibition. Despite its significance, we know little about its structure.

Until recently, troponin resisted crystallization attempts, and the troponin complex is too big for present NMR techniques. Of course, one can ask why we need the structure. Crystallographers promised us years ago that if we only knew the structure we would know everything about the function. Well, that's not quite so. However, hype notwithstanding, working without a structure is like working in a darkened room. For example, we might know that certain parts of a molecule move, and other parts change the orientation or move away from each other. Even if we know where our probe is and what residues it labels, we will not know which part of the molecule is which. Molecular structure is not everything, but without it a lot of experiments are reduced to fingerprinting with no meaningful interpretation. In the case of the regulatory complex of muscle consisting of TnI – inhibitory subunit, TnC – Ca sensor, tropomyosin, and TnT that anchors troponin to tropomyosin and actin, the structure of some components is known (Figure 10). TnC was crystallized and its structure solved nearly 20 years ago. Tropomyosin has been crystallized, but it does not diffract well, preventing the determination of a high-resolution structure. TnI and TnT have not been crystallized, although there is a preliminary but unpublished report of the crystal structure of the complex. Moreover, from a number of biochemical and biophysical studies, it seems almost a given that the structure of the isolated components is not the same as their structure within the complex. Thus, before we could ask any meaningful questions about function, we need to know the structure of the complex.

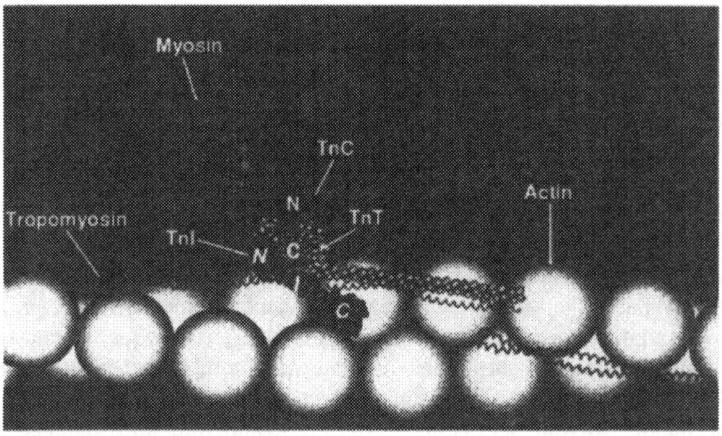

Figure 10. A model of the troponin regulatory complex of muscle. TnI—inhibitory subunit, TnC—Ca sensor, tropomyosin, and TnT, which anchors troponin to tropomyosin and actin (courtesy of J. Potter).

The conformational changes that constitute Ca sensing have been implied by comparison of the Ca-free and Ca-bound structures of isolated TnC (Figure 11). During Ca binding to the regulatory sites in the N-terminal lobe of TnC, two helixes rotate and expose a hydrophobic patch. I call this an "armpit model": calcium binds, and the arm lifts up exposing a surface to which another protein can bind (Herzberg and James, 1985). It has been postulated that the other protein is TnI. In the absence of Ca, the troponin complex interacts with tropomyosin or actin. The binding of Ca to TnC exposes the hydrophobic patch, and TnI binds to the hydrophobic patch reducing its interactions with tropomyosin and actin and somehow relieving the inhibition of myosin ATPase. To validate this model, we need to verify the movement of TnI and its location relative to TnC and TnT.

Currently, there are two competing models that show what the TnI/TnC binary complex looks like; both are computational. The first one, a helical model from Maeda's group (Vassylyev et al., 1998), was built by homology to a structure of a similar

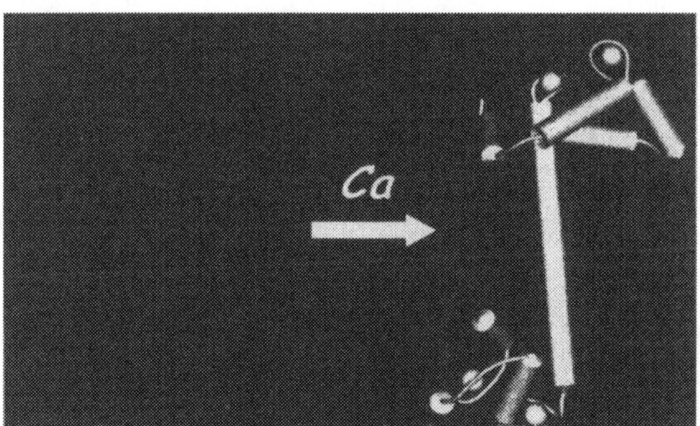

Figure 11. X-tal structures of Ca-free and Ca-bound TnC. The two halves of the molecule, N-domain (top) and C-domain (bottom), are very similar. Note reorientation of helices in the N-domain.

N-terminal peptide of TnI complexed with the C-lobe of TnC (Figure 12). TnC is a rather symmetric molecule with a similar structure of the C- and N-terminal lobes. It was therefore reasonable to assume that a similar peptide interacting with a C-lobe will interact in an analogous manner with the N-lobe of TnC (Figure 12, yellow). The second competing model, on the left, is the beta-hairpin model of Trehwella and collaborators (Tung et al., 2000). This particular model was built on the basis of neutron scattering, NMR data of small fragments, and homology of gelsolin-actin interactions. Instead of a helix, the same region is a beta-hairpin.

We have here a controversy between the x-ray crystallographers and NMR spectroscopists.What better technique to settle the dispute than EPR? We jumped into the fray, employing the site-directed spin-labeling (SDSL) technique that was developed originally by Wayne Hubbell and Christian Altenbach in the late 1980s (Altenbach et al., 1989). SDSL is a combination of cysteine scanning and the application of three different EPR measurements: solvent accessibility, mobility, and interspin distance (Figure 13).

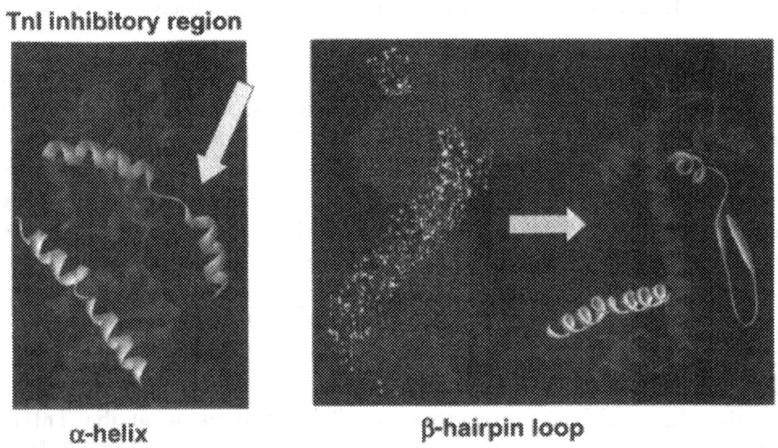

Figure 12. Competing computational models of TnI/TnC binary complexes. Left: Helical model developed by Maeda et al (Vassylyev et al., 1998). Right: Beta-hairpin model of Trewhella et al. (Tung et al., 2000)

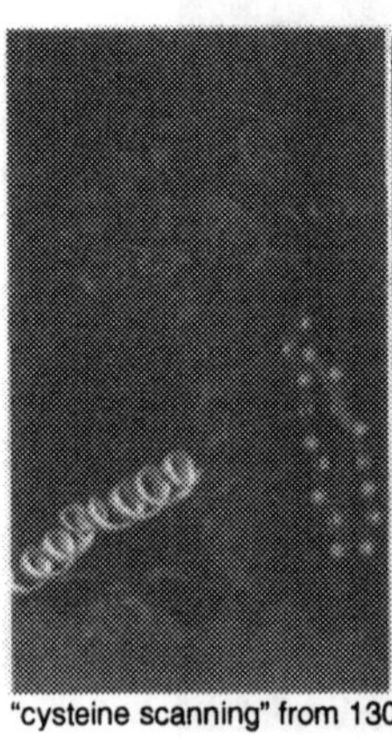

Spectroscopy	Molecular property	Signal	
Power saturation	Solvent accessibility	Amplitude	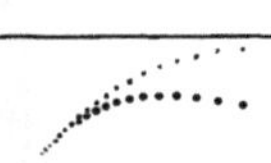
Conventional EPR	Mobility	Splitting	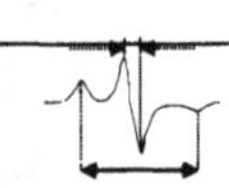
Dipolar EPR	Spin-spin distance	Broadening	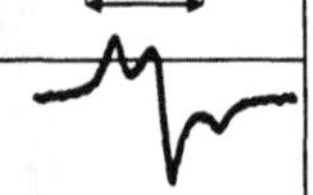

Figure 13. The experimental overview of the SDSL-EPR used to identify the structure of binary complexes TnI and TnC. See Figure 6 for the cysteine scanning image.

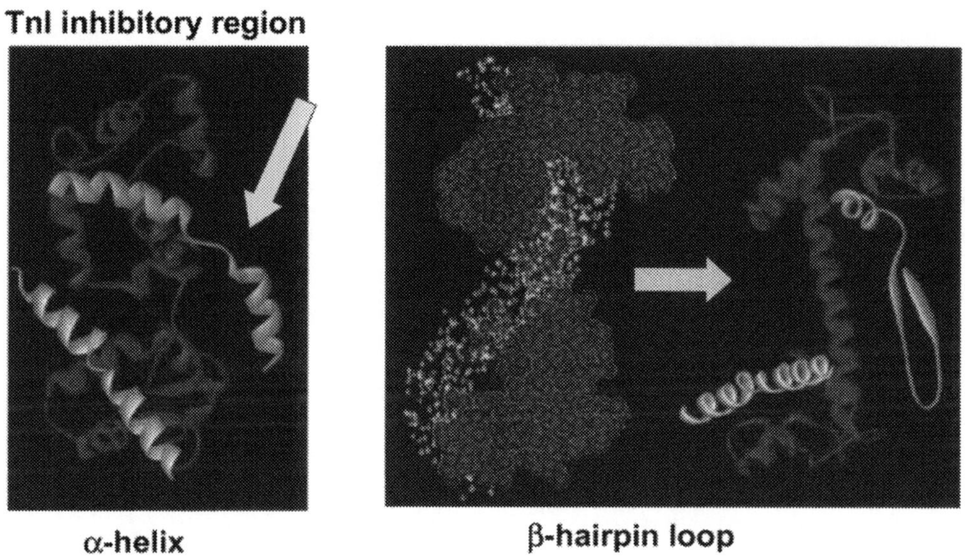

Figure 12. Competing computational models of TnI/TnC binary complexes. Left: Helical model developed by Maeda et al (Vassylyev et al., 1998). Right: Beta-hairpin model of Trewhella et al. (Tung et al., 2000)

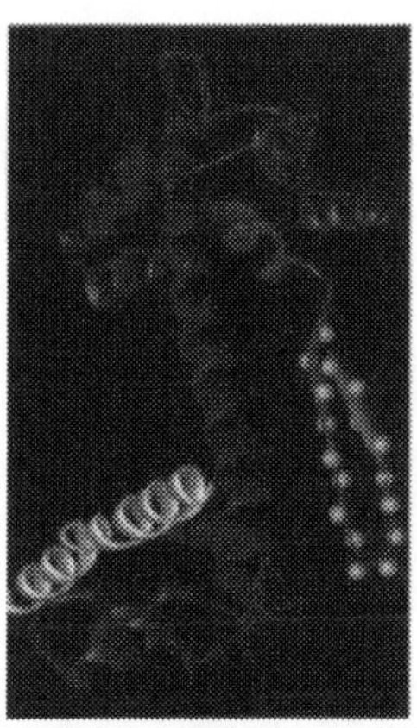

"cysteine scanning" from 130-146

Spectroscopy	Molecular property	Signal	
Power saturation	Solvent accessibility	Amplitude	
Conventional EPR	Mobility	Splitting	
Dipolar EPR	Spin-spin distance	Broadening	

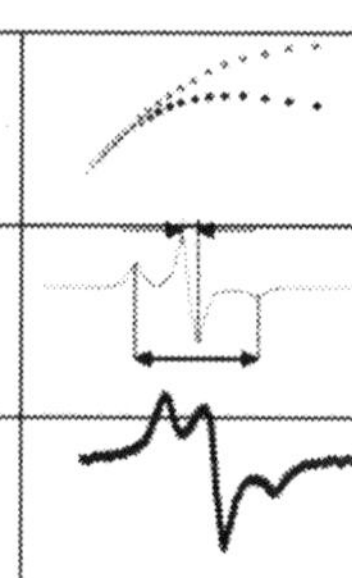

Figure 13. The experimental overview of the SDSL-EPR used to identify the structure of binary complexes TnI and TnC. See Figure 6 for the cysteine scanning image.

We measure these parameters for each residue in turn and look for a periodic trend. If the periodicity is 2 residues, the secondary structure is a beta strand; if the periodicity is 3.6 residues, the structure is an alpha helix. We did precisely these kinds of measurements, "we" being an extremely good postdoctoral research associate, Louise Brown, and her subgroup of a couple of undergraduate students who created 20 mutants in 3 months. I still don't know how they did it, but I don't want to ask too many questions (Brown et al., 2002).

Figure 14 shows the profile of solvent accessibility in the binary complex of TnC and TnI. There is no trend along the sequence of 18 residues, not the 3.6 periodicity e) pected from helical model, nor the periodicity of 2 residues of the beta hairpin model. This could mean (1) we don't know what we are doing, or (2) the inhibitory region is disordered. However, when we added a third component, TnT, to the troponin, we obtained a beautiful pattern of solvent accessibility with a periodicity of 3.6 residues. Figure 15 compares the experimental data with the computational predictions for the solvent accessibility for the two models. It is clear that the helical model fits the experiment, while the beta-hairpin does not. When we move along the sequence towards the C-terminus the periodicity is lost (Figure 16), and that sequence stretch of residues is disordered. This was somewhat expected, as some isoforms of TnI have two consecutive prolines in the region wrecking the helix. This stretch of 6-8 residues has been implicated by FRET and crosslinking experiments to be a hinge that might allow the C-terminal domain of TnI to move between the hydrophobic path of TnC and actin while anchored at the N-terminal end to TnC.

Comparison of the binary (TnC*I) and the ternary complex (+TnT) data reveals the position of TnT. The mobility of each residue is sensitive to tertiary interactions. If a label is placed on a surface of the helix facing the solvent it will move fast. This fast movement averages the anisotropy of the signal and results in a sharp signal (Figure 17, left side). If the label is trapped between the surfaces of two subunits, its motion will be sterically hindered, and the spectrum will be broad. We use this simple principle to find the interface between TnI and TnT. The right side of Figure 17 maps the mobility changes induced by addition of TnT onto the surface of TnI. Such mapping identifies the residues from that interface; they are in the N-terminus of the scanned sequence.

In summary, using SDSL-EPR we have found that: (a) the N-terminus of the inhibitory sequence has a helical structure stabilized by interactions with TnT, and (b) the helix is followed by a disordered region that might be a switch. Preliminary reports of the crystal structure of the ternary complex support our findings. In this crystal structure, the scanned sequence forms a coiled-coil with TnT at the N-terminus, then there is a break in

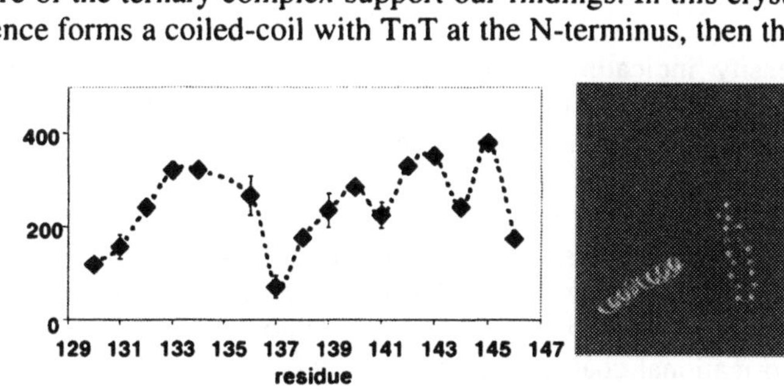

Figure 14. Solvent accessibility of the TnC and TnI binary complexes along the inhibitory region of TnI. See Figure 6 for the cysteine scanning image.

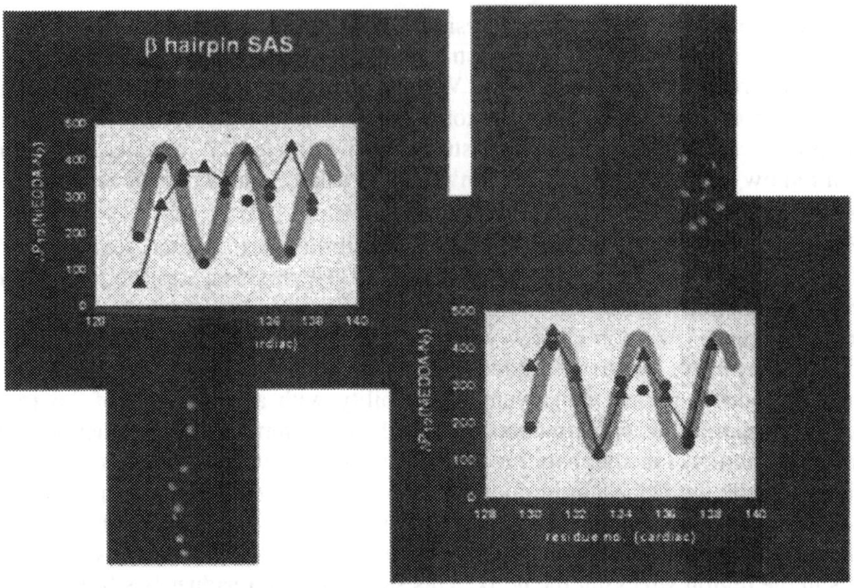

Figure 15. Solvent accessibility of the TnC, TnI and TnT ternary complexes between residues 129 and 138 (*circles*), the labeled residues are shown for the helical and beta-hairpin models. Top: Comparison of experimental data with computational predictions (*triangles*) based in the helical model. Bottom: Beta-hairpin model showing alpha-helical periodicity *(gray line)*.

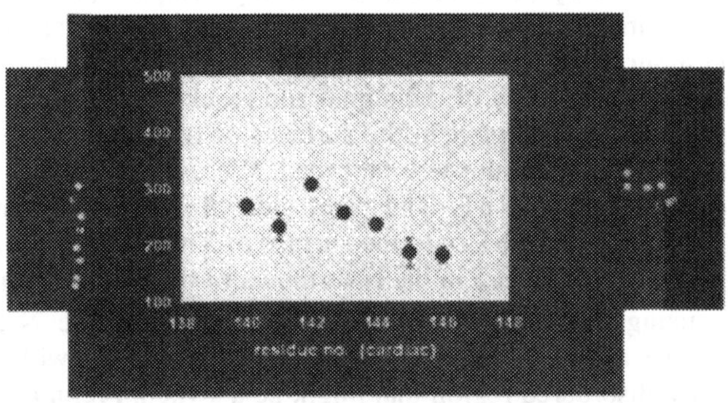

Figure 16. Solvent accessibility for residues 139-146. Note, there is no periodicity.

the electron density indicating disorder, which we see also in solution. This work illustrates complementarity between the low-resolution structure by EPR and high-resolution structure by x-ray or maybe even NMR. We are not competing with these methods; rather, we want to take advantage of the selectivity of EPR to move to a higher level to determine the organization of these complexes. As mentioned before, all of these studies can be performed in muscle fibers facilitating the correlation of function and force development, something that X-ray or NMR cannot do. We have now more than 40 mutants of TnC and TnI that we are exchanging into muscle fibers and correlating the kinetics of conformational changes as reflected by orientation, distance, and the kinetics of force generation. To describe the conformational changes is not good enough; we have

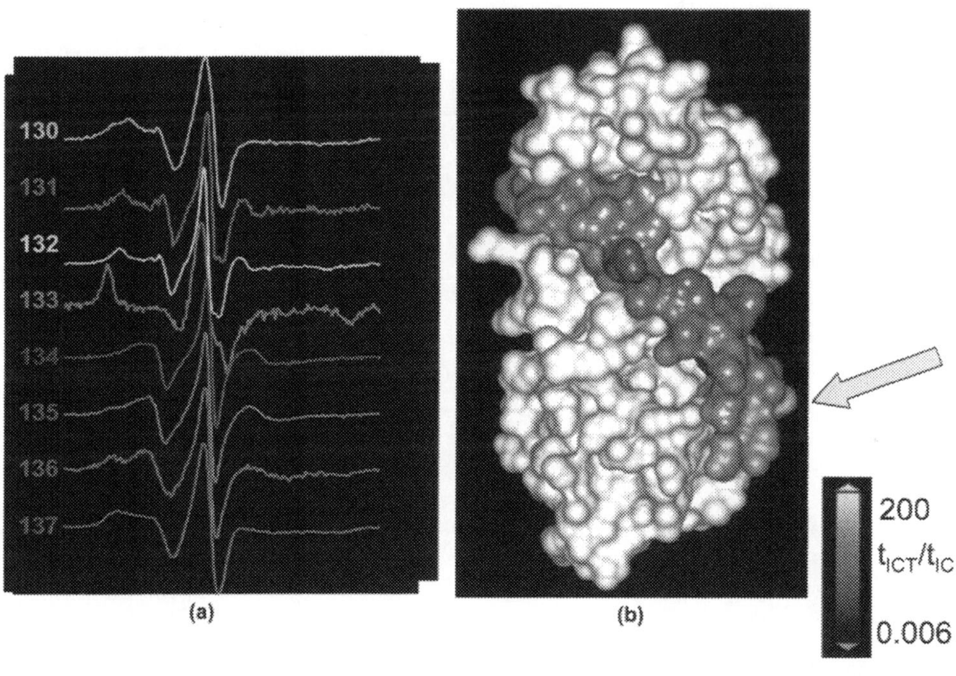

Figure 17. Surface residue mobility comparison of the binary (TnC*I) and ternary complex (+TnT) data reveals the position of TnT. (a) Residue mobility is determined from the spectra. Addition of TnT broadens the spectra of residues 130-136 (not shown), identifying the interface between TnI and TnT; (b) coloring of the TnI residues reflects changes in the rate of motion between ternary and binary complexes. *Yellow* indicates decrease of motion. *Blue* indicates increase of motion.

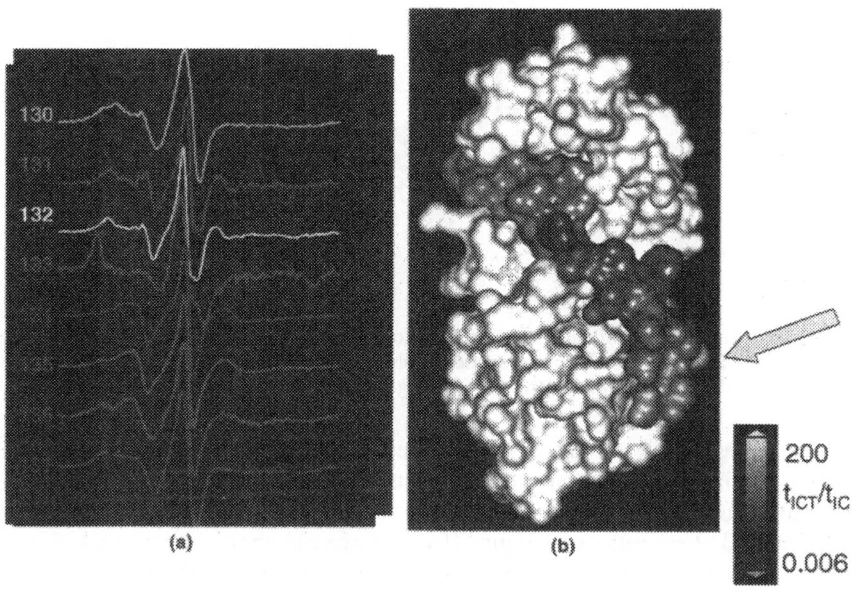

Figure 17. Surface residue mobility comparison of the binary (TnC*I) and ternary complex (+TnT) data reveals the position of TnT. (a) Residue mobility is determined from the spectra. Addition of TnT broadens the spectra of residues 130-136 (not shown), identifying the interface between TnI and TnT; (b) coloring of the TnI residues reflects changes in the rate of motion between ternary and binary complexes. *Yellow* indicates decrease of motion. *Blue* indicates increase of motion.

to prove that these conformational changes cause or accompany the function of the protein. One way to do this is to correlate the structure and function in the time domain.

6.4. ORIENTATION AND DYNAMICS OF THE MYOSIN HEAD

The other part of the biological motor complex is of course the engine itself—myosin. The myosin engine works by bringing two ends of the muscle cell together. The muscle cell consists of a series of smaller subunits: sarcomeres made of actin filaments; the Z-line that delineates the sarcomere at either end and to which the actin filaments attach; and the myosin filaments in the middle of the sarcomere that overlap the actin. Hugh Huxley and Andrew Huxley (unrelated researchers), independently of each other postulated that the contraction takes place by the filaments sliding past each other and pulling actin filaments from opposite sides of the sarcomere towards the middle, thus shortening the length of the muscle cell (Figure 18) (Huxley and Hanson, 1954; Huxley and Niedergerke, 1954).

Later, Hugh Huxley proposed that it's probably the reorientation of the myosin head when attached to the actin filament that pulls on actin (Huxley, 1969). This model is generally accepted and is in all the textbooks. For the past 40 years, however, many labs using a variety of techniques have searched for proof of myosin head reorientation, and the final verdict is still out.

Our approach to finding the reorientation of the myosin head is to describe the orientation and dynamics of the myosin head before and after the force is generated. We

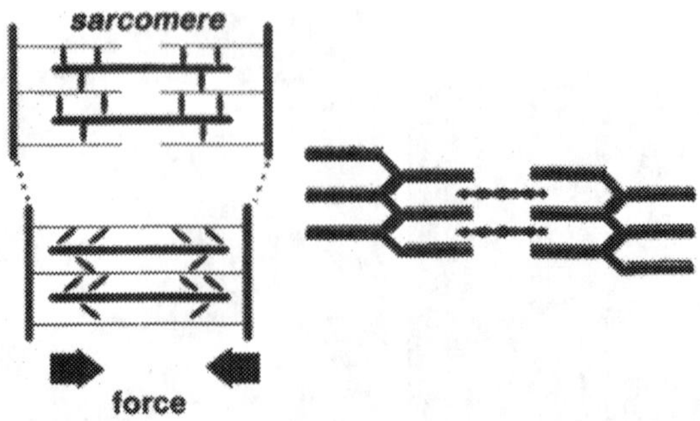

Figure 18. Sliding filament model of muscle contraction. Rotation of myosin heads bridging actin and myosin filaments induces the strain between them that is relieved when the filaments slide relative to each other. Since actin filaments are anchored to the end of sarcomere, sliding towards the middle shortens the muscle.

dissect the ATPase cycle by trapping intermediate states such as the detached (from actin) state, initial weakly attached state, strongly attached state before the force generated, after the force generated, and before and after the products of ATP hydrolysis, phosphate, and ADP are released (Figure 19). The general approach is seen in Figures 20 and 21. We trap the cycle in a given state, collect the spectra in the oriented muscle

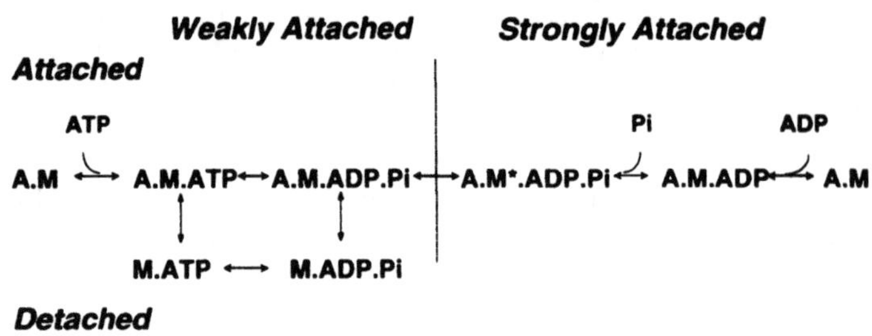

Figure 19. Intermediate states of myosin ATPase cycle. The strength of attachment of the head to the actin filaments increases from left to right as the ATP is hydrolyzed and products are released.

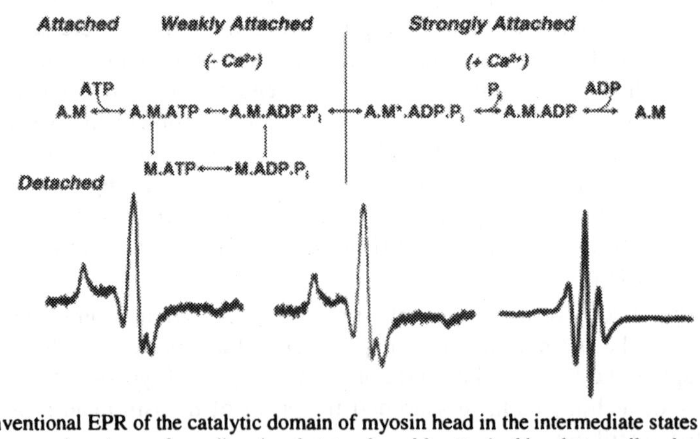

Figure 20. Conventional EPR of the catalytic domain of myosin head in the intermediate states of the ATPase cycle indicate progressive change from disordered state of weakly attached heads to well-ordered heads after product release.

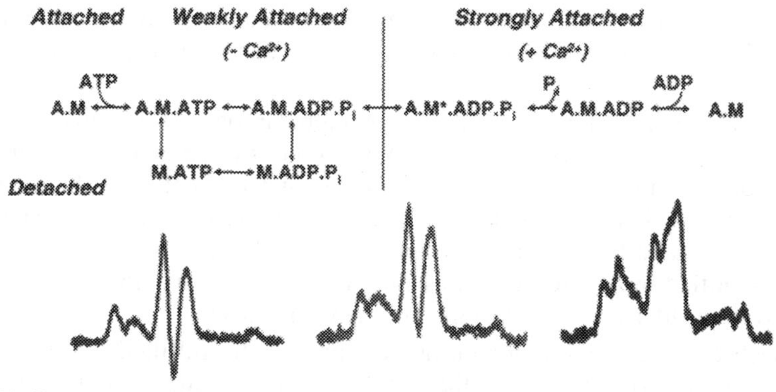

Figure 21. ST-EPR spectra of the muscle fibers indicate decreasing mobility of the catalytic domain.

fibers, determine by EPR the orientational distribution, and determine by ST-EPR the dynamics of the head in each state (Raucher and Fajer, 1994; Raucher et al., 1994; Fajer et al., 1991).

On the basis of such measurements, we proposed the disorder-to-order model for the force generation shown in Figure 22 (Raucher and Fajer, 1994). Detached heads are completely disordered and mobile. Initial docking of the heads on the actin filament is not stereo-specific and (surprisingly) leaves the head still fairly mobile. As the ATP is hydrolyzed, but before the products of hydrolysis are released, the mobility decreases, but the attachment is still not stereo-specific. After the force is generated and the first hydrolysis product, phosphate, is released from the active site, the head is rigidly attached to actin (on the millisecond timescale) and is also extremely well ordered. This change from disordered-mobile myosin heads to ordered-rigid heads is what we think generates the force.

In molecular terms, disorder might represent an initial formation of an interface between actin and myosin. Formation of stereo-specific contacts between actin and myosin forces myosin into a single orientation, straining the rest of the molecule. This strain is the muscle force. If the ends of the sarcomere are released, the strain is relieved by the sliding of the filaments past each other. Energetically this is quite feasible; the energy released by the actin-myosin complex during formation of this stereo-specific interface is larger than the energy of ATP hydrolysis. It can easily drive the mechanical work. This is not such a revolutionary concept as it seems—Andrew Huxley proposed

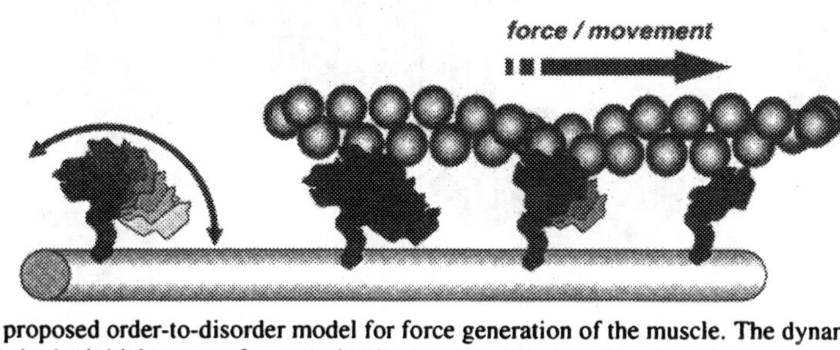

Figure 22. The proposed order-to-disorder model for force generation of the muscle. The dynamically disordered heads in the initial stages of contraction become more immobilized and better ordered as hydrolysis products are released. The ordering of the heads results in the change of average orientation.

this model nearly 30 years ago. It is as elegant an explanation as that of the rotation of the myosin head from one angle before the force is generated to another angle to generate the force.

This model was proposed for the large globular part of the myosin head-motor domain. At the same time, the crystal structure of the myosin head was solved revealing two large domains: the motor domain and the elongated, light-chain-binding regulatory domain (Figure 23). Ivan Rayment, who solved the structure, immediately postulated that the strain is generated by the rearrangement of some elements within the motor domain, and then magnified by a lever arm consisting of the regulatory domain (Rayment et al., 1993). It was the latter that was rotating, not the whole head.

For such a model, two structural aspects have to be fulfilled: (1) there must be a hinge between the two domains so that the lever arm can move; and (2) the lever arm itself must be stiff to transmit the force to the myosin filaments. To test both aspects we compared the mobility of the motor domain and regulatory domain and the mobility of the distal and proximal parts of the regulatory domain. The rationale was that if there is a hinge, then the motion of different domain will be uncoupled; if the domain is stiff, it will move as a rigid body with the same mobility along the entire structure.

We placed spin and phosphorescent labels on the motor domain, essential light chain (proximal part of the regulatory domain), and regulatory light chain (distal part of the regulatory domain) (Figure 24). The spectra are shown in Figure 24 and are summarized in the model shown in Figure 25. We found that the regulatory domain moves with nearly the same rate as the essential light chain; thus the domain moves as a rigid body and it is stiff (Baumann et al., 2001). The catalytic domain, on the other hand, moves three times faster than the regulatory domain, revealing a putative hinge (Adhikari et al., 1997). Because ST-EPR does not measure the amplitude of motion, we measured the amplitudes with phosphorescent anisotropy (Figure 26). The residual anisotropy a few milliseconds after the excitation that selects one orientation of probes is related to the amplitude of motions. If the protein does not move, the initial and final anisotropies will be the same,

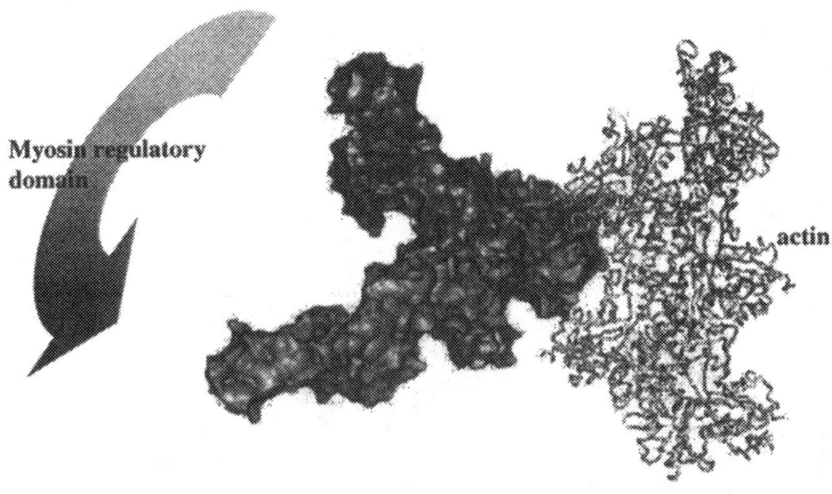

Figure 23. The comparison of different crystal structures of the myosin head suggests the rotation of the light-chain-binding regulatory domain.

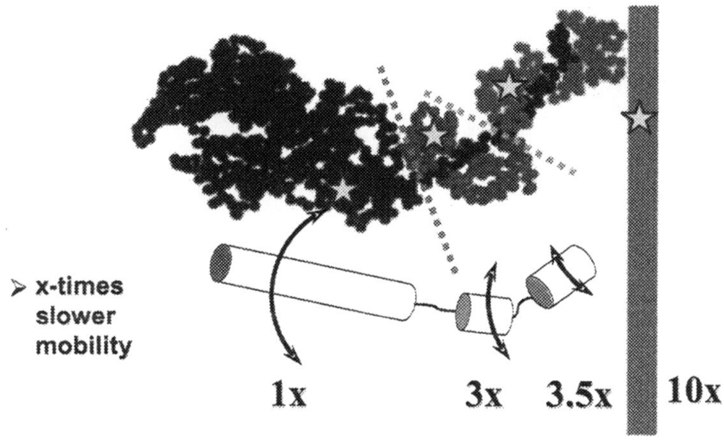

Figure 25. Intradomain mobility within myosin head, showing the motor domain (*blue*), essential light chain (*red*), and the regulatory light chain (*green*) and filament surface (*grey*).

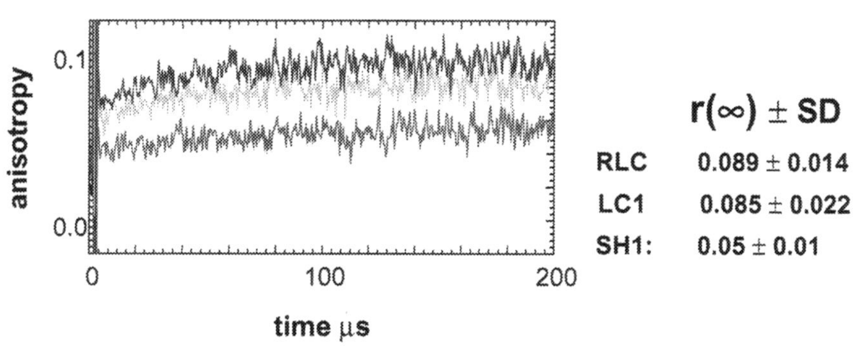

Figure 26. Time-resolved phosphorescence anisotropy revealing the amplitudes of motion, catalytic domain (*red*), essential light chain (*yellow*), and regulatory light chain (*purple*).

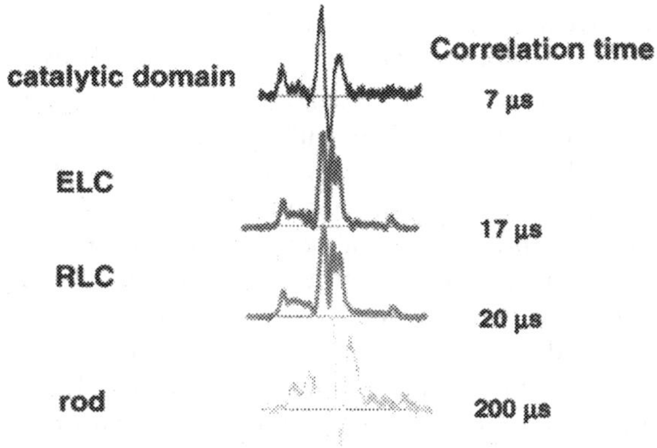

Figure 24. ST-EPR spectra of the motor (catalytic) domain and the regulatory domain of myosin head in synthetic myosin filaments. The essential light chain is the proximal part of the regulatory domain; the regulatory light chain is the distal part of the regulatory domain. For comparison, the mobility of the filament core (rod) is shown.

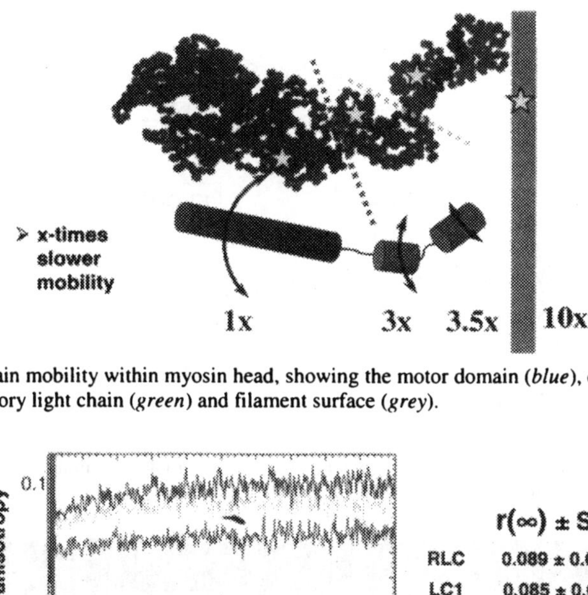

Figure 25. Intradomain mobility within myosin head, showing the motor domain (*blue*), essential light chain (*red*), and the regulatory light chain (*green*) and filament surface (*grey*).

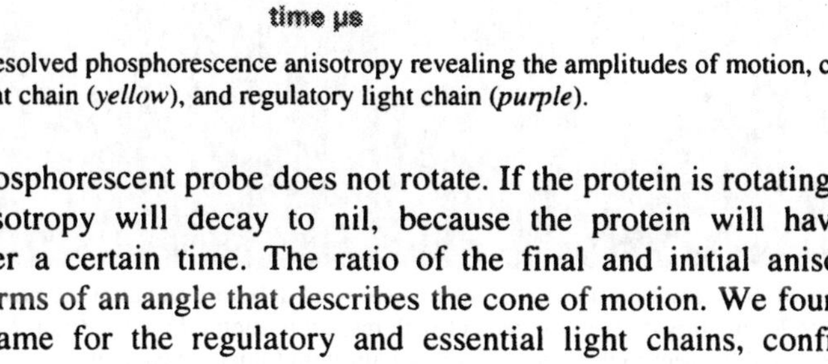

Figure 26. Time-resolved phosphorescence anisotropy revealing the amplitudes of motion, catalytic domain (*red*), essential light chain (*yellow*), and regulatory light chain (*purple*).

because the phosphorescent probe does not rotate. If the protein is rotating fast and freely, the initial anisotropy will decay to nil, because the protein will have no preferred orientation after a certain time. The ratio of the final and initial anisotropies can be expressed in terms of an angle that describes the cone of motion. We found that the cone angle is the same for the regulatory and essential light chains, confirming that the regulatory domain is a rigid body. The cone angle for the catalytic domain was twice as big, again consistent with the hinge between the two domains (Brown et al., 2001).

These studies tell us a lot about the dynamics of the two domains but not about their relative geometry. This we accomplished with FRET. We placed fluorescent pairs of donor and acceptor molecules in the motor and regulatory domains and measured distances between them using frequency-modulated FRET (Figure 27) to obtain the lifetimes of the donor in the presence and absence of the acceptor molecule. The increase of relaxation rate of the donor in the presence of acceptor can be related to the distance between the two. The result was pretty much as expected; molecular modeling of the distance we measured and the distances measured by others were consistent with the large reorientation of the regulatory domain in various intermediate states of the ATPase cycle. The red area in Figure 28 denotes positions of the end of the regulatory domain before and after the power stroke and are consistent with measured distance (Palm et al., 1999). The head can change its shape consistent with the rotation of the regulatory domain.

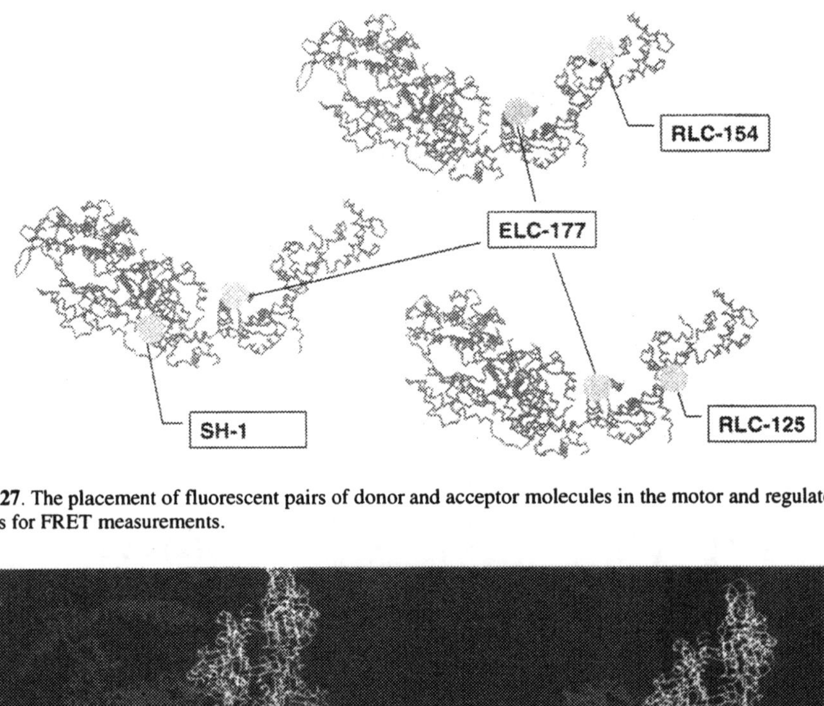

Figure 27. The placement of fluorescent pairs of donor and acceptor molecules in the motor and regulatory domains for FRET measurements.

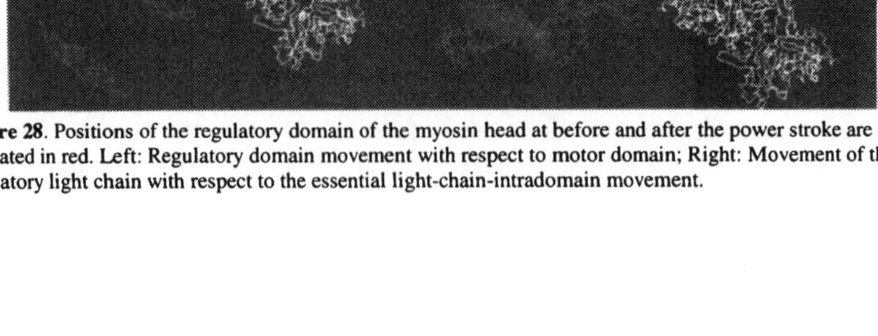

Figure 28. Positions of the regulatory domain of the myosin head at before and after the power stroke are indicated in red. Left: Regulatory domain movement with respect to motor domain; Right: Movement of the regulatory light chain with respect to the essential light-chain-intradomain movement.

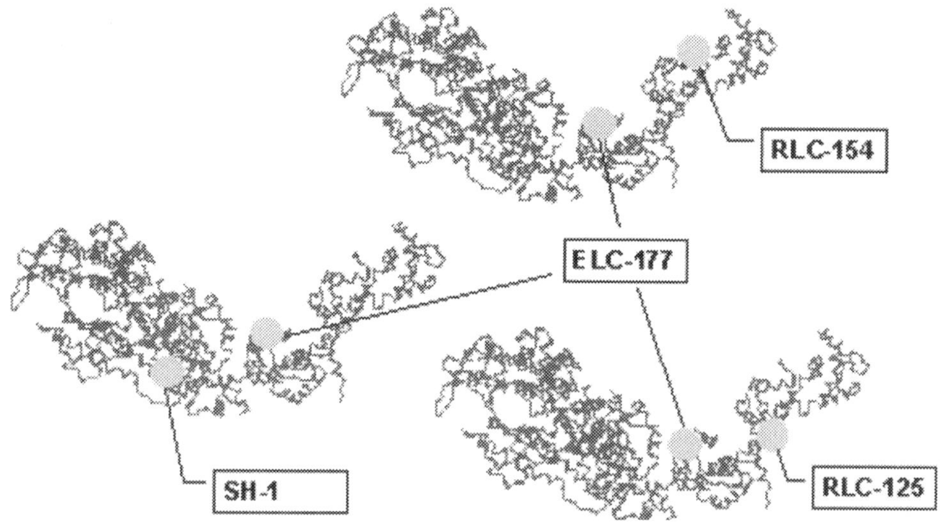

Figure 27. The placement of fluorescent pairs of donor and acceptor molecules in the motor and regulatory domains for FRET measurements.

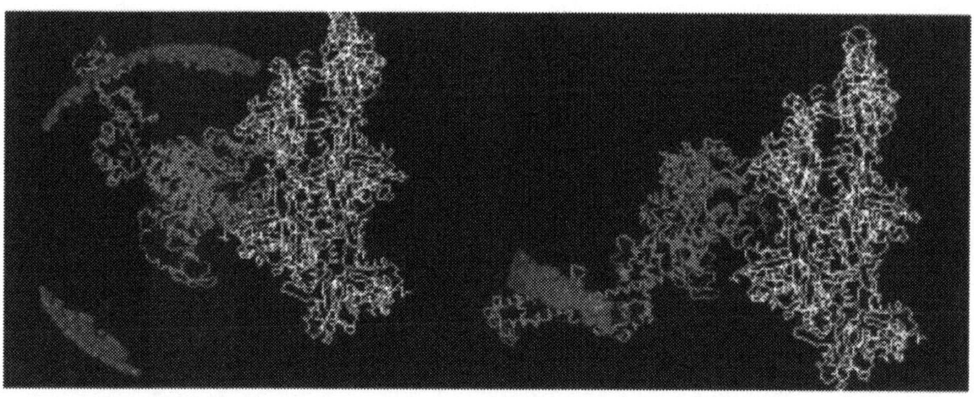

Figure 28. Positions of the regulatory domain of the myosin head at before and after the power stroke are indicated in red. Left: Regulatory domain movement with respect to motor domain; Right: Movement of the regulatory light chain with respect to the essential light-chain-intradomain movement.

Now when we look at the distances within the regulatory domain, there is no movement in the axial direction that can move myosin relative to actin. The only movement that would be consistent with the FRET measurement is a perpendicular, azimuthal motion that can add flexibility to the myosin head but can't account for muscle force.

In summary, all three methods—ST-EPR, phosphorescence anisotropy, and FRET—reveal (a) a hinge between the motor and regulatory domains, and (b) the stiff regulatory domain. This means that the structure of the myosin head is capable of generating translating conformational changes within the motor domain to the swing of the regulatory domain, and that the regulatory domain is rigid enough to act as a lever arm.

All these experiments were performed using myosin filament in solution in the absence of actin. The challenge now is to catch these conformational changes in *flagrante delicto.*

6.5. ACKNOWLEDGMENTS

I was blessed with a number of excellent students and postdocs. Louise Brown came just for a year and a half to my lab and taught us molecular biology as well as spearheading the troponin effort. She was in charge of the group of two undergraduates, Ron Hills and Clement Rouviere; and two first-year graduate students, Xiaojun Zhang and Likai Song. Our computational guys, Ken Sale, now at Sandia National Laboratories, was our main molecular modeler and Khaled Khairy, now at Max Planck Institute in Dresden, was instrumental in the development of EPR simulation programs. Bruce Baumann, Bishow Adhikari, and postdocs Thomas Palm and Huichun Li worked on the dynamics of myosin. Outside collaborators were Brett Hambly and Bill Sawyer of the University of Sydney and the University of Melbourne, respectively. Kalman Hideg synthesized the spin labels we use. Our corporate sponsors are the Muscular Dystrophy Association, the American Heart Foundation, and the National Science Foundation.

6.6. REFERENCES

Adhikari, B., Hideg, K., and Fajer, P. G., 1997, Independent mobility of catalytic and regulatory domains of myosin heads, *Proc. Natl. Acad. Sci. USA* **94**(18):9643-9647.

Altenbach, C., Flitsch, S. L., Khorana, H. G., and Hubbell, W. L., 1989, Structural studies on transmembrane proteins. 2. Spin labeling of bacteriorhodopsin mutants at unique cysteines, *Biochemistry* **28**(19)7806-7612.

Baumann, B. A., Hambly, B. D., Hideg, K., and Fajer, P. G., 2001, The regulatory domain of the myosin head behaves as a rigid lever, *Biochemistry* **40**(26):7868-7873.

Brown, L. J., Klonis, N., Sawyer, W. H., Fajer, P. G., and Hambly, B. D., 2001, Independent movement of the regulatory and catalytic domains of myosin heads revealed by phosphorescence anisotropy, *Biochemistry* **40**(28):8283-8291.

Brown, L. J., Hills, R., Rouviere, C., Sale, K., and Fajer, P., 2002, Structure of the inhibitory region of cardiac TnI in the ternary complex by SDSL-EPR spectroscopy, *Biochem. J.* **82**(1):830 Part 2.

Fajer, P. G., Fajer, E. A., Schoenberg, M., and Thomas, D. D., 1991, Orientational disorder and motion of weakly attached cross-bridges, *Biophys. J.* **60**(3):642-649.

Herzberg, O., and James, M. N., 1985, Structure of the calcium regulatory muscle protein troponin-C at 2.8 A resolution, *Nature* **313**(6004):653-659.

Huxley, H. E., 1969, The mechanism of muscular contraction, *Science* **164**(886):1356-1365.

Huxley, H. E., and Hanson, J., 1954, Changes in the cross-striations of muscle during contraction and stretch and their structural interpretation, *Nature* **173**:973-976.

Huxley, A. F., and Niedergerke, R., 1954, Structural changes in muscle during contraction. Interference microscopy of living muscle fibres, *Nature*, **173**:971-973.

Jeschke, G., 2002, Distance measurements in the nanometer range by pulse EPR, *Chemphyschem* **3**(11):927-932.

Milov, A. D., Mar'yasov, A.G., Samoilova, R. I., Tsvetkov, Yu D., Raap, J., Monaco, V., Formaggio, F., Crisma, M., and Toniolo, 2000, Biochemistry, biophysics, and molecular biology-pulsed electron double resonance of spin-labeled peptides: data on the peptide-chain secondary structure, *Dokl. Biochem.* **370**(1-6):8-11.

Palm, T., Sale, K., Brown, L., Li, H. C., Hambly, B., and Fajer, P. G., 1999, Intradomain distances in the regulatory domain of the myosin head in prepower and postpower stroke states: fluorescence energy transfer, *Biochemistry* **38**(40):13026-13034.

Rabenstein, M. D., and Shin, Y. K., 1995, Determination of the distance between two spin labels attached to a macromolecule, *Proc. Natl. Acad. Sci. USA* **92**(18):8239-8243.

Rayment, I., Rypiewski, W. R., Schmidtbase, K., Smith, R., Tomchick, D. R., Bemming, M. M., Wunkelmann, D. A., Wesenberg, G., and Holden, H. M., 1993, Three-dimensional structure of myosin subfragment-1: a molecular motor, *Science* **261**(5117):50-58.

Raucher, D., and Fajer, P. G., 1994, Orientation and dynamics of myosin heads in aluminum fluoride induced pre-power stroke states: an EPR study, *Biochemistry* **33**(39):11993-11999.

Raucher, D., Sar, C. P., Hideg, K., and Fajer, P. G., 1994, Myosin catalytic domain flexibility in MgADP, *Biochemistry* **33**(47):14317-14323.

Sale, K., Sar, C., Sharp, K. A., Hideg, K., and Fajer, P. G., 2002, Structural determination of spin label immobilization and orientation: a monte carlo minimization approach, *J. Magn. Reson.* **156**(1)104-112.

Tung, C. S., Wall, M. E., Gallagher, S. C., and rewhella, J., 2000, A model of troponin-I in complex with troponin-C using hybrid experimental data: the inhibitory region is a beta-hairpin, *Protein Science* **9**(7):1312-1326.

Vassylyev, D. G., Takeda, S., Wakatsuki, S., Maeda, K., and Maeda, Y., 1998, Crystal structure of troponin C in complex with troponin I fragment at 2.3-A resolution, *Proc. Natl. Acad. Sci. USA* **95**(9)4847-4852.

GENE REGULATORY NETWORKS

7. THE INTRICATE WORKINGS OF A BACTERIAL EPIGENETIC SWITCH

Aaron Hernday, Bruce Braaten, and David Low*

ABSTRACT

Bacteria have developed epigenetic mechanisms to control the reversible Off-to-On switching of cell surface structures such as pyelonephritis-associated pili (PAP). The *pap* pili switch is primarily controlled by the global regulator leucine-responsive regulatory protein (Lrp), the local regulator PapI, and DNA adenine methylase (Dam). There are two sets of binding sites for Lrp in the *pap* regulatory region: promoter proximal sites 1,2,3 and promoter distal sites 4,5,6. The pilin promoter proximal ($GATC^{prox}$) and distal ($GATC^{dist}$) targets for Dam are located within Lrp binding sites 2 and 5, respectively. In the Off state, Lrp binds cooperatively to sites 1,2,3 overlapping the *papBA* pilin promoter, shutting off pilin transcription, and blocking methylation of $GATC^{prox}$. Binding of Lrp at sites 1,2,3, together with methylation of $GATC^{dist}$, reduces the affinity of Lrp for sites 4,5,6, preventing simultaneous binding of Lrp at sites 4,5,6 upstream. Switching to the phase On state requires the environmentally regulated PapI co-regulator, which increases the affinity of Lrp for sites 5 and 2. PapI binds specifically to Lrp-*pap* DNA complexes via binding with Lrp as well as contact with DNA sequences within *pap* sites 5 and 2. Directionality in switching from Off to On appears to be due to methylation of $GATC^{prox}$, which prevents formation of the PapI-Lrp-*pap* site 2 ternary complex. A switch model is presented in which DNA replication is proposed to play a critical role by generating a hemimethylated $GATC^{dist}$ site and displacing Lrp from sites 1,2,3. This facilitates methylation of $GATC^{prox}$ and binding of PapI-Lrp to sites 4,5,6, with subsequent activation of *pap* transcription. The first gene product of the *pap* operon, PapB, positively regulates *papI* transcription, resulting in a positive feedback loop that helps maintain the On state. The *pap* switch is environmentally regulated by a number of factors including the CpxAR two-component regulatory system, the Histone-like nucleoid structuring protein H-NS, and cAMP-Catabolite Gene Activator Protein (CAP), which all involve binding of regulatory binding proteins to *pap* DNA sequences with subsequent alteration of PapI and Lrp binding. The Pap switch mechanism, with

---

* Aaron Hernday, Bruce Braaten, and David Low, University of California, MCD Biology, Santa Barbara, CA 93106.

interesting variations, is conserved among a number of enteric bacteria, controlling expression of many unrelated pili-adhesin complexes.

7.1. INTRODUCTION

DNA methylation patterns control gene expression in diverse organisms from bacteria to humans (Casadesus and Torreblanca, 1996; Hendrich and Bird, 2000). In eukaryotes methylation at cytosine in 5′-CG-3′ DNA sequence inhibits gene expression by the binding of proteins known as methyl cytosine binding proteins (MeCPs), which then attract histone deacetylases that silence gene expression (Nan et al., 1998). In bacteria DNA methylation patterns have been shown to play a more complex role in controlling individual gene expression (Braaten et al., 1994; van der Woude et al., 1996). The focus of this seminar is on the mechanisms by which DNA methylation controls phase variation of Pap in uropathogenic *Escherichia coli* (UPEC). Phase variation is defined as the switching On and Off of gene expression. In the bacterial world there are a number of phase variation mechanisms, most of which involve changes in the DNA sequence such as inversions, insertions, or deletions of regulatory DNA's (Henderson et al., 1999). The *pap* operon differs from these other systems since it is controlled by DNA methylation, and thus does not involve changes in the DNA sequence.

7.2. MODEL

The core *pap* switch requires the global regulator leucine-responsive regulatory protein (Lrp), the local regulator PapI, and Dam (Hernday et al., 2002). In the Pap phase Off state (transcriptionally inactive), Lrp binds cooperatively to sites 1,2,3 overlapping the *papBA* pilin promoter (Figure 1). The switch to the On state is facilitated by PapI, which is required for movement of Lrp from sites 1,2,3 to sites 4,5,6 over 100 base-pairs upstream of the *papBA* promoter. PapI is a small (8 kilodaltons) regulatory protein that does not bind to *pap* regulatory DNA with any measurable affinity (Kaltenbach et al., 1995). It binds, however, to both Lrp and DNA sequences within *pap* sites 2 and 5 that enable formation of stable PapI-Lrp-*pap* DNA ternary complexes at both pilin promoter distal sites 4,5,6 as well as proximal sites 1,2,3. Methylation of the promoter proximal GATC site within Lrp binding site 2 (denoted $GATC^{prox}$) is essential for transition to the phase On state. Methylation of $GATC^{prox}$ blocks PapI-dependent formation of Lrp-PapI-*pap* ternary complexes at sites 1,2,3, facilitating movement of Lrp/PapI to distal sites 4,5,6.

We hypothesize that transition to the On state requires DNA replication to displace Lrp from *pap* sites 1,2,3, allowing $GATC^{prox}$ to be methylated which blocks PapI-dependent binding of Lrp to sites 1,2,3. In addition, we hypothesize that PapI-Lrp binding to sites 4,5,6 is facilitated by the hemimethylated state of $GATC^{dist}$ immediately following DNA replication. That is, the affinity of PapI-Lrp for sites 4,5,6 containing hemimethylated $GATC^{dist}$ should be higher than for sites 4,5,6 containing a fully methylated $GATC^{dist}$. Preliminary data support this hypothesis.

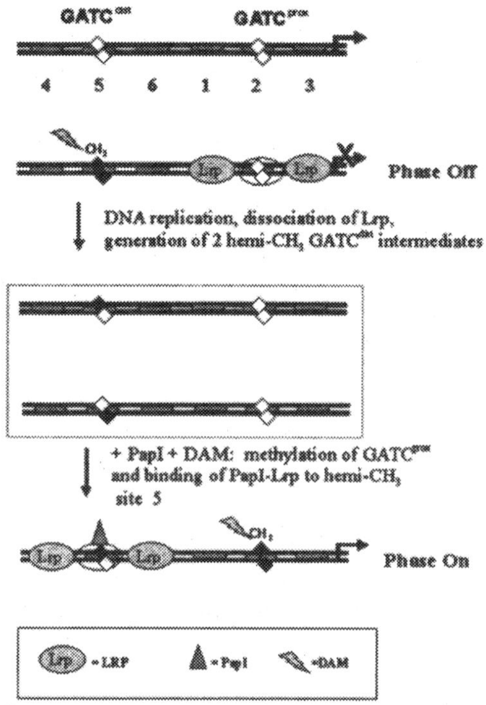

Figure 1. Pap phase variation model: Off to On switch.

Binding of Lrp/PapI at sites 4,5,6, together with binding of cAMP-CAP 60 bp (66.5bp from site 5 and 34.5bp from site 4) upstream of Lrp binding site 4, stimulates transcription at the *papBA* pilin promoter (Weyand et al., 2001). The first gene of this operon, *papB*, encodes a regulatory protein that binds with highest affinity to a site near the divergent *papI* promoter, and stimulates *papI* transcription. This constitutes a positive feedback loop for phase On cells, maintaining the phase On state by increasing levels of PapI necessary for binding of Lrp to the distal sites 4,5,6.

7.2.1. Data Supporting the Core Switch Model

The balance between phase Off and phase On states is controlled by the competition between Dam, PapI, and Lrp binding at *pap* sites 1,2,3 and sites 4,5,6. Initial observation showed that the wild-type switch is Off in a *dam* null mutant background as shown in Figure 2 (Braaten et al., 1994). This occurs since under conditions in which methylation of $GATC^{prox}$ is blocked, PapI-Lrp binds to sites 1,2,3 and inhibits *papBA* transcription (Weyand and Low, 2000). Further analysis showed that overproduction of Dam (by 4-fold and higher) also blocked *pap* transcription, locking cells in the phase Off state (Braaten et al., 1994). Why is this so? Based on the Pap phase variation model shown in Figure 1, we reason that immediately following DNA replication there is a window of time during which the $GATC^{dist}$ site remains hemimethylated until Dam binds to and methylates the $GATC^{dist}$ site. There are normally only about 100 molecules of Dam in the cell, just sufficient to methylate the 20,000 or so GATC sites in the *E. coli* chromosome in one cell cycle (Boye et al., 1992). Under conditions of Dam overproduction, Dam will

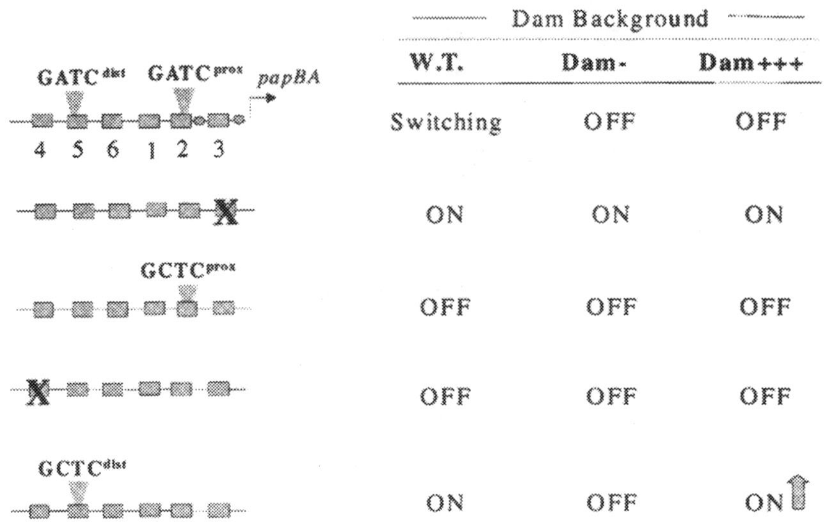

Figure 2. Effects of alteration of Dam level and *pap* regulatory sequences on Pap phase variation. "ON" denotes a phase locked ON switch phenotype, "OFF" denotes a phase locked OFF switch phenotype, "switching" denotes the wild-type, reversible OFF-ON switch phenotype, and "X" indicates a mutation in an Lrp binding site.

compete more effectively with PapI-Lrp for binding at $GATC^{dist}$, preventing transition to the phase On state. This hypothesis is based on the assumption that binding of PapI-Lrp to *pap* sites 4,5,6 containing a fully methylated $GATC^{dist}$ sequence is inhibited significantly more than binding to hemimethylated $GATC^{dist}$, which needs to be tested.

In the phase Off state Lrp is bound to sites 1,2,3, blocking methylation of $GATC^{prox}$, and $GATC^{dist}$ is unoccupied and fully methylated. This state is stabilized by methylation of $GATC^{dist}$, which reduces the affinity of Lrp for sites 4,5,6. In addition, binding of Lrp at sites 1,2,3 reduces the affinity of Lrp for sites 4,5,6 by 10-fold via a phenomenon that we have called "mutual exclusion," which requires that the DNA be negatively supercoiled as it naturally occurs in *E.coli* (Hernday et al., 2002). One way that this might occur is via bending of DNA by Lrp at sites 1,2,3, resulting in a distortion of the DNA helix near sites 4,5,6, altering spacing between the Lrp binding sites. This could reduce Lrp's affinity for sites 4,5,6 by reducing cooperativity.

Mutation of Lrp binding site 3 results in a phase-locked On phenotype shown in Figure 2, which is independent of PapI and Dam (Nou et al., 1995). This observation is consistent with the model since this mutation reduces the affinity of Lrp for *pap* sites 1,2,3, thereby reducing the (negative) mutual exclusion effect on binding of Lrp to sites 4,5,6. Under these conditions the affinity of Lrp for sites 4,5,6 increases, obviating the need for PapI and Dam (Figure 2).

For the phase ON state to form, PapI-dependent binding of Lrp at *pap* site 5 must occur together with Lrp binding at sites 4 and 6. Mutation of Lrp binding site 4 blocks *pap* transcription by inhibiting binding of Lrp to site 4 (Figure 2). In contrast, mutation of $GATC^{dist}$ to $GCTC^{dist}$ locks the switch in the phase ON state by blocking methylation of $GATC^{dist}$ without significantly affecting binding of PapI-Lrp (Braaten et al., 1994). Under these conditions binding of Lrp to sites 4,5,6 is no longer inhibited by methylation

(Figure 2). However, in the absence of Dam the GCTC^{dist} mutant is locked Off since methylation of GATC^{prox} is still required to prevent binding of PapI-Lrp at sites 1,2,3.

A systems model of the core *pap* phase switch will require inclusion of the interactions shown in Figure 3. This includes competition between Lrp and Dam at GATC^{dist} and PapI-Lrp and Dam at GATC^{prox}. Validity and usefulness of any model(s) will require that one can predict the switch outcome of altering parameters such as Dam, PapI and Lrp levels. Our preliminary results show that as Dam levels are raised above "0," Off to On switching increases to a maximum, and then as Dam levels get even higher switching decreases to "0" (not shown). The positive slope (ascending to the maximum switch rate as Dam levels increase) is greater than the negative slope (descending to zero at even higher Dam levels), reflecting the different competitions occurring at the *pap* sites. The ascending curve represents methylation at GATC^{prox}, which has a positive effect on switching. The descending curve represents methylation at GATC^{dist}, which has a negative effect on switching (see Figure 1).

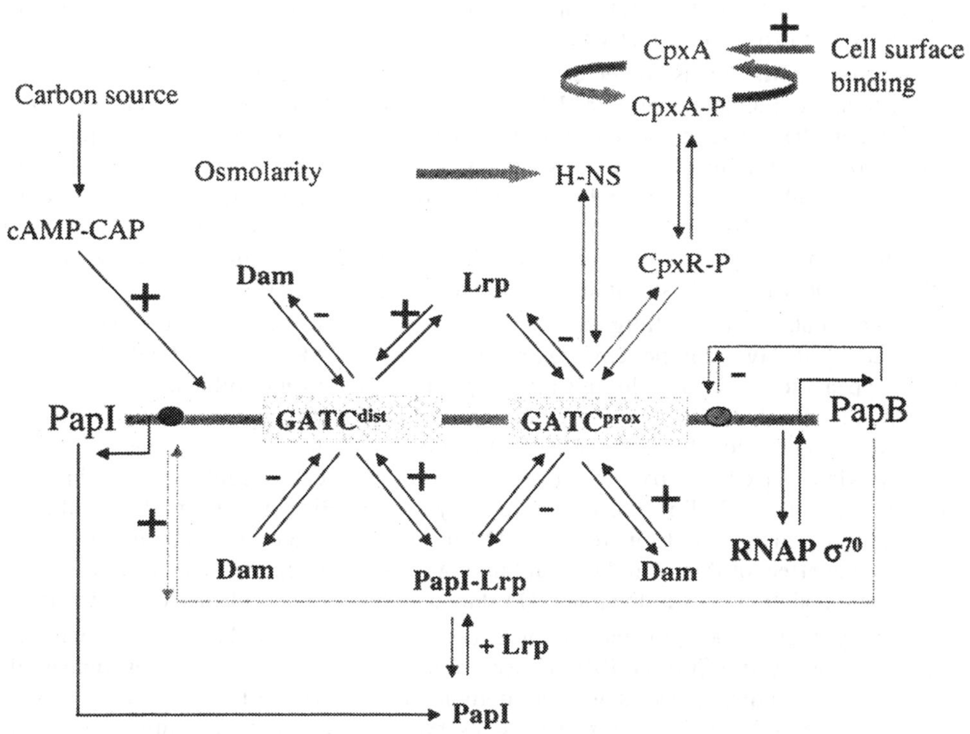

Figure 3. Pap switch parameters.

7.3. BEYOND THE CORE SWITCH: ENVIRONMENTAL EFFECTS ON PAP PHASE VARIATION

Binding of PapI-Lrp to site 5 and Lrp to sites 4 and 6 is required for *pap* transcription (Nou et al., 1995). In addition, binding of catabolite gene activator protein (CAP) is required at a site upstream of *pap* site 5 (Goransson et al., 1989). Nathan Weyand showed that the *papBA* promoter proximal subunit of CAP is required to activate *pap* transcription, likely via interaction with the α-C-terminal domain of RNA polymerase (RNAP) (Weyand et al., 2001). Although the CAP binding site of *pap* is located 215.5 bp from the transcription start of the *papBA* promoter, the interaction between CAP and RNAP likely occurs via a large loop since activation is helical phase-dependent (Weyand et al., 2001). The role of Lrp binding at sites 4,5,6 is not yet clear. Lrp could bend the DNA and facilitate cAMP-CAP-RNAP interaction, or could contact RNAP itself. Notably, Lrp alone is sufficient to activate *pap* transcription in vitro using purified RNAP. Addition of cAMP-CAP further increases *pap* transcription in the presence of Lrp, but is unable to activate transcription in the absence of Lrp. These results suggest the possibility that Lrp might directly interact with RNAP.

Besides cAMP levels, other environmental factors influence Pap phase variation. The histone-like regulatory protein H-NS influences the Off to ON switch rate since in an *hns*-deficient background the Pap Off to On switch rates are reduced (White-Ziegler et al., 2000). This indicates that H-NS helps activate Pap phase switching, which may occur as a result of specific binding of H-NS to *pap* sequence near $GATC^{prox}$ (White-Ziegler et al., 2000), and competition with binding of Lrp to sites 1,2,3. H-NS is regulated by a variety of environmental conditions including pH and osmolarity, and thus links Pap expression to these conditions (White-Ziegler et al., 2000).

Recent data from S. Hultgren and T. Silhavy indicate that Pap expression is also regulated by the two-component regulatory system CpxAR (Otto and Silhavy, 2002). CpxA appears to sense the folding states of proteins such as *pap* pilin subunits, and may provide a feedback system by which binding of Pap pili to receptors on eukaryotic cells transmit a signal to the bacterium. Activation of CpxA in the inner membrane phosphorylates CpxR to CpxR-P, which binds to a number of regulatory sites in the *E. coli* chromosome including the *pap* regulatory region. Binding of CpxR-P to the *pap* regulatory region appears to increase the Off to On switch rate by competing with PapI-Lrp. The net effect of this would be that binding of UPEC to epithelial cells in the urinary tract would maintain cells in the phase On state and thus further enhance colonization.

Finally, many other pili operons in *E. coli* and *Salmonella* share features with *pap* including conserved ACGATCTTT sequences in *pap* sites 2 and 5, which encompass the Lrp and PapI binding sites as well as homologues to the PapI co-regulatory protein. Notably, some of these pili operons have conserved CAP binding sites similar to *pap* that would tie expression of these pili to cAMP levels. Others, such as the *pef* operon in *Salmonella*, lack a CAP binding site. Instead, Pef pili phase variation is acid-controlled. At pH 7 very few cells express Pef pili whereas at pH 4.5 transcription is turned on to a high level based on fluorescence microscopy using anti-Pef pili antisera (Nicholson and Low, 2000). This sensor system may allow *Salmonella* entering the stomach to switch Pef pili on, which are then used in the intestines to facilitate binding of *Salmonella* and contribute to colonization and pathogenesis (Baumler et al., 1996).

7.4. ACKNOWLEDGMENTS

Thanks to all former Low Lab members with special thanks to Marjan van der Woude, Linda Kaltenbach, Nathan Weyand, and Margareta Krabbe. We thank the National Institutes of Health for continued support of this project (AI23348).

7.5. REFERENCES

Baumler, A. J., Tsolis, R. M., Bowe, F. A., Kusters, J. G., Hoffmann, S., and Heffron, F., 1996, The *pef* fimbrial operon of Salmonella typhimurium mediates adhesion to murine small intestine and is necessary for fluid accumulation in the infant mouse, *Infect. Immun.* **64**:61-68.

Boye, E., Marinus, M. G., and Lobner-Olesen, A., 1992, Quantitation of Dam methyltransferase in *Escherichia coli*, *J. Bacteriol.* **174**:1682-1685.

Braaten, B. A., Nou, X., Kaltenbach, L. S., and Low, D. A., 1994, Methylation patterns in pap regulatory DNA control pyelonephritis-associated pili phase variation in E. coli, *Cell* **76**:577-588.

Casadesus, P., and Torreblanca, J., 1996, Methylation-related epigenetic signals in bacterial DNA, in *Epigenetic Mechanisms of Gene Regulation*, V. E. A. Russo, R. A. Martienssen, and A. D. Riggs, eds., Cold Spring Harbor Laboratory, New York, pp. 141-153.

Goransson, M., Forsman, P., Nilsson, P., and Uhlin, B. E., 1989, Upstream activating sequences that are shared by two divergently transcribed operons mediate cAMP-CRP regulation of pilus-adhesin in *Escherichia coli*, *Mol. Microbiol.* **3**:1557-1565.

Henderson, I. R., Owen, P., and Nataro, J. P., 1999, Molecular switches—the ON and OFF of bacterial phase variation, *Mol. Microbiol.* **33**:919-932.

Hendrich, B., and Bird, A., 2000, Mammalian methyltransferases and methyl-CpG-binding domains: proteins involved in DNA methylation, *Curr. Top. Microbiol. Immunol.* **249**:55-74.

Hernday, A., Krabbe, M., Braaten, B., and Low, D., 2002, Self-perpetuating epigenetic pili switches in bacteria, *Proc. Natl. Acad. Sci. USA* **99**(4):16470-16476.

Kaltenbach, L. S., Braaten, B. A., and Low, D. A., 1995, Specific binding of PapI to Lrp-pap DNA complexes, *J. Bacteriol.* **177**:6449-6455.

Nan, X., Cross, S., and Bird, A., 1998, Gene silencing by methyl-CpG-binding proteins, *Novartis Found Symp.* **214**:6-16.

Nicholson, B., and Low, D., 2000, DNA methylation-dependent regulation of pef expression in *Salmonella typhimurium*, *Mol. Microbiol.* **35**:728-742.

Nou, X., Braaten, B., Kaltenbach, L., and Low, D. A., 1995, Differential binding of Lrp to two sets of pap DNA binding sites mediated by Pap I regulates Pap phase variation in Escherichia coli, *EMBO J* **14**:5785-5797.

Otto, K., and Silhavy, T. J., 2002, Surface sensing and adhesion of Escherichia coli controlled by the Cpx-signaling pathway, *Proc. Natl. Acad. Sci. USA* **99**:2287-2292.

van der Woude, M., Braaten, B., and Low, D., 1996, Epigenetic phase variation of the pap operon in *Escherichia coli*, *Trends Microbiol.* **4**:5-9.

Weyand, N. J., and Low, D. A., 2000, Regulation of Pap phase variation. Lrp is sufficient for the establishment of the phase off *pap* DNA methylation pattern and repression of *pap* transcription in vitro, *J. Biol. Chem.* **275**:3192-3200.

Weyand, N. J., Braaten, B. A., van der Woude, M., Tucker, J., and Low, D. A., 2001, The essential role of the promoter-proximal subunit of CAP in pap phase variation: Lrp- and helical phase-dependent activation of *papBA* transcription by CAP from -215, *Mol. Microbiol.* **39**:1504-1522.

White-Ziegler, C. A., Villapakkam, A., Ronaszeki, K., and Young, S., 2000, H-NS controls *pap* and *daa* fimbrial transcription in *Escherichia coli* in response to multiple environmental cues, *J. Bacteriol.* **182**:6391-6400.

8. YEAST SIGNAL TRANSDUCTION: REGULATION AND INTERFACE WITH CELL BIOLOGY

George F. Sprague; Paul J. Cullen, and April S. Goehring*

ABSTRACT

We examined the morphogenetic transitions that yeast cells undergo in response to extracellular cues, and determined that multiple mechanisms control specificity of signal transduction pathway signaling and the attendant physiological response that ensues. This article describes the approaches that we used to determine these mechanisms. Our findings indicate that scaffolding proteins, which organize signal transduction pathways, are an especially powerful means to achieve specificity. We do not yet know how general this mechanism is. Our studies have also started to reveal ways in which a protein, Ste20, first identified as a participant in signal transduction pathways, may also connect to the basic cell biology machinery. Synthetic lethal genetic analysis has suggested that the polarisome and a new ubiquitin-like system may be targets of Ste20.

8.1. INTRODUCTION

Yeast cells can undergo morphogenetic transitions in response to extracellular cues (Figure 1). A classic example is the change from bud initiation to shmoo formation that occurs when the mating pheromone response pathway is activated by binding of pheromone to its cell surface receptor (Figure 1, lower left). A second morphogenetic transition, discovered in Gerry Fink's lab at MIT about a decade ago, occurs in response to nutrient limitation, and is the result of activation of a distinct signal transduction pathway, the filamentous growth pathway. The cells change their growth habit dramatically. In the case of vegetative growth (abundant nutrients), the cells are spherical and have elaborated new buds close to their birth site, producing a tight cluster of cells (Figure 1, top). This budding pattern is termed the axial pattern. In response to nutrient limitation, the cells exhibit a different budding pattern: they bud from the pole opposite

* George F. Sprague, Paul J. Cullen, and April S. Goehring, Institute of Molecular Biology, University of Oregon, Eugene, OR 97403.

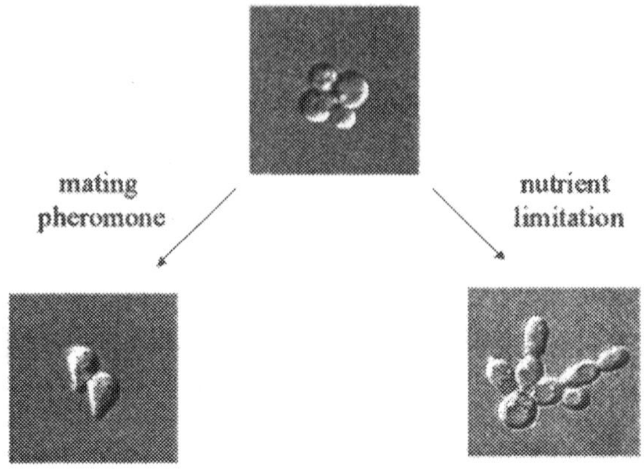

Figure 1. Yeast morphogenetic transitions. A microcolony of yeast cells growing vegetatively in the presence of abundant nutrients is shown in the top panel. In the panel on the lower left, shmoos that result from treatment of cells with mating pheromone are shown. In the panel on the right, a microcolony undergoing filamentous growth in response to nutrient limitation is shown.

their birth site (Figure 1, lower right). They also change their shape: they are considerably elongated compared to the vegetative cells. Hence, nutrient limitation and the concomitant switch to filamentous growth can be thought of as a foraging behavior (Gimeno et al., 1992). The cell has recognized that nutrients are limiting and changed its growth habit so that its progeny can sample new environments.

Although the pheromone response pathway and the filamentous growth pathway are distinct and elicit distinct physiological changes, studies over the last half-a-dozen years have led to the surprising realization that the two pathways share a number of protein components. This realization naturally leads to the question of how pathway specificity is maintained. For example, if a particular protein that participates in both pathways has been activated by the pheromone pathway, what prevents it from also causing stimulation of downstream components in the filamentous growth pathway? In this article, we will explore mechanisms that confer specificity to pathway signaling and we will also explore the interface of these signaling pathways with basic cell biology machinery, interfaces that enable these pathways to elicit the appropriate morphogenetic and physiological changes.

8.2. SIGNALING SPECIFICITY

The pheromone response pathway, sometimes called the mating pathway, serves as an excellent entrée for the discussion of the mechanisms that govern the specificity of signal transduction pathway activity. In the pathway shown in Figure 2, each of the two yeast mating types, **a** and α, secretes a small peptide pheromone that can bind to a receptor present at the surface of the other mating type. These receptors are members of the serpentine 7 transmembrane family and couple to a trimeric G protein. Activation of the G protein in turn activates a quartet of protein kinases that act sequentially: Ste20, Ste11, Ste7, and Fus3. This MAP kinase cascade influences the activity of a transcription factor and, thereby, increases the transcription of genes whose products catalyze the

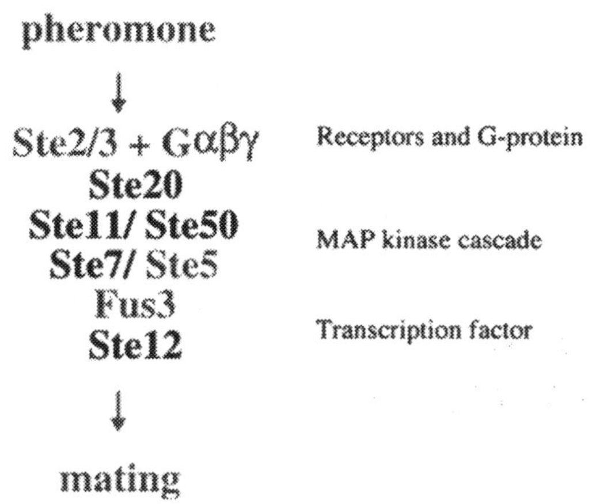

Figure 2. The pheromone response pathway. The components of the pheromone response pathway are indicated. Of particular note are the proteins Ste20, Ste11 and Ste7, which are components of a MAP kinase cascade used in other signal transduction pathways; Ste5 is thought to be a scaffolding protein that functions to confer specificity.

actual mating event. (Parenthetically, and for completeness' sake, we note that activation of this pathway regulates two other target proteins in addition to the transcription factor. First, Fus3 controls the activity of a cyclin-dependent kinase inhibitor, Far1, and in this way contributes to arrest of the mitotic cell division cycle. Second, the activated G protein influences the activity of Cdc42, a p21 GTPase that is required to establish subcellular polarity and orient the actin cytoskeleton. The appropriate activation of Cdc42 is required for shmoo formation.)

The filamentous growth pathway uses many components that were first identified as part of the pheromone response pathway (Figure 3). Notably, elements of the MAP kinase cascade—Ste20, Ste11 and Ste7—are shared by mating and filamentous pathways. The filamentous pathway does appear to use a unique terminal MAP kinase, Kss1. Hence, a more explicit formulation of the specificity question would be: How do Ste20, Ste11, and Ste7 know in which pathway they are participating? The problem becomes even more interesting with the realization that a third signal transduction pathway, one that is activated in response to high osmolarity of the growth medium, also uses these components of the MAP kinase cascade (Figure 3). Moreover, a second specificity question arises. At the top of the filamentous and osmolarity pathways, there are no known components that distinguish one from the other; both rely on the membrane protein Sho1. We will explore each of these specificity issues in turn.

Scaffolds: One solution to the specificity problem appears to come from scaffolding proteins, proteins that bind to each member of the MAP kinase cascade and, thereby, organize it into a complex. For example, in the case of the pheromone response pathway, Ste5 has been shown to bind to Ste11, Ste7, and Fus3, and also to a component of the trimeric G protein. Hence, it is reasonable to posit that Ste5 organizes the pheromone response pathway components into a multi-component complex and thereby insulates the pheromone MAP kinase cascade from other MAP kinase cascades present in the yeast cell. A clear prediction of this point of view is that in a cell lacking the Ste5 protein, activation of the pheromone response pathway should lead to spillover or bleedthrough to

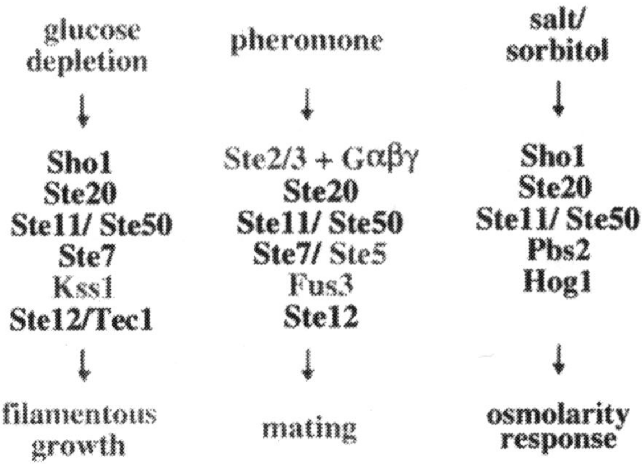

Figure 3. Signaling pathways share components. The components of the filamentous growth, mating, and osmolarity response pathways are shown.

other MAP kinase cascades. In my laboratory, John Printen, in collaboration with Beverly Errede's laboratory, tested this prediction (Yashar et al., 1995). Their experiment is shown in Figure 4 in cartoon fashion. In a wild-type cell, Ste5 acts as a scaffold and organizes the pathway. Hence, when the cell is stimulated with pheromone, the signal passes through the pathway and there is the appropriate transcription readout. However, in a *ste5* mutant, activation of the pheromone pathway leads to bleedthrough to other MAP kinase cascades. Printen and Errede happened to examine readout of the Pkc1 MAP kinase cascade, one not mentioned yet, but nonetheless the important point is that bleedthrough to that pathway was seen in the *ste5* mutant.

Hiten Madhani and colleagues performed a similar experiment and demonstrated that in *ste5* mutants, activation of the pheromone pathway led to bleedthrough of the signal to the filamentous growth pathway, resulting in transcription activation of targets of the

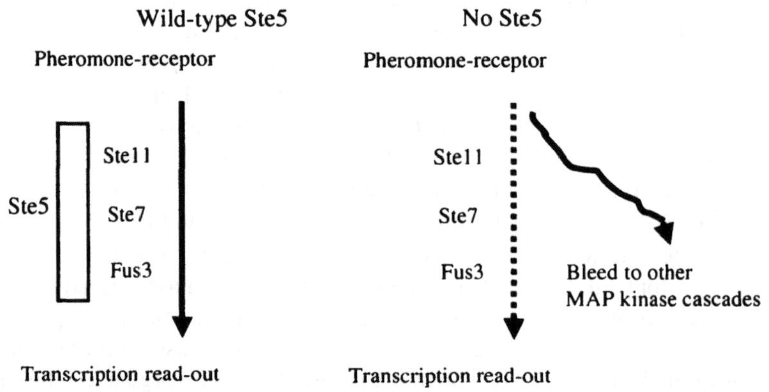

Figure 4. Scaffolds as specificity factors. The expected consequences for signal transmission in the presence and absence of Ste5 are illustrated. In a wild-type cell, Ste5 is present and the signal generated by the activated pheromone-receptor complex passes through the pheromone response MAP kinase cascade to give the appropriate transcription readout. In the absence of Ste5, some of the signal can bleed to other MAP kinase cascades present in the cell.

filamentous growth pathway (H. Madhani, personal communication). Hence, in *ste5* mutants, the pheromone signal gets "lost" and activates other signal transduction pathways.

The osmolarity pathway also appears to maintain specificity through the activity of a scaffold protein. In this case, there is an interesting twist: the scaffold is Pbs2, which is the central element of the three-tiered osmolarity MAP kinase cascade (Posas and Saito, 1997). O'Rourke and Herskowitz showed that activation of the osmolarity pathway in *pbs2* mutants led to bleedthrough to the pheromone pathway (O'Rourke and Herskowitz, 1998). Together then, these experiments lead to the general notion that scaffolding proteins may organize signal transduction pathways into multi-component complexes, and thereby ensure that the initial stimulus is automatically tied to the appropriate transcription readout by virtue of that complex.

Heteromeric receptors: A second specificity mechanism has been revealed by the effort to understand how the filamentous and osmolarity pathways may be distinguished at the very top. As already noted, both pathways require the Sho1 protein, a presumptive membrane protein that may serve as a receptor. Alternatively, Sho1 may sense properties of the plasma membrane itself and thereby lead to activation of downstream pathways. Are the filamentous and osmolarity pathways really identical at the top, as they now seem? We entertained the hypothesis that there were as-yet-unidentified components that would act at the membrane and confer specificity. We performed microarray analysis to identify genes whose transcription was both Ste12-dependent and -induced under filamentous growth conditions. A surprisingly small number of genes met this pair of criteria. We have pursued one, *MSB2*, a gene that had been previously identified but whose role in the yeast cell was poorly understood. This gene is particularly interesting because it encodes a presumptive membrane protein and, hence, might be a missing specificity factor. Indeed, not only is *MSB2* a transcription target of the filamentous growth pathway, but the product is itself required for filamentous growth. An *msb2* mutant cannot undergo filamentous growth and, moreover, overexpression of the *MSB2* product leads to enhanced filamentous growth. Before presenting those data, a side trip to illustrate more fully the filamentous growth response and the assays used to measure it is appropriate.

Figure 5 is a cartoon of the axial growth pattern that was presented in the photographs of real cells in Figure 1 (top). The mother cell has made a bud, which has grown and separated from the mother following mitosis. Both the mother and that first bud have now initiated a new budding cycle, producing buds that are next to the site where the first bud emerged–the axial pattern. In filamentous growth, the first bud has emerged from the mother cell, but note that it is considerably elongated compared to the mother by the time it has completed its mitotic cell cycle and separated from the mother cell. In the next budding cycle, its first bud emerges at the pole opposite from its birth point, a pattern termed "unipolar budding." Hence, microscopic examination is one convenient assay for filamentous growth. A second assay is simply to wash the surface of the Petri dish after colonies have formed on the plate. If filamentous growth has occurred (after about 2 days), some cells have invaded the agar and a scar of the colony is apparent (Figure 6). The invasion is apparent in a cross-section of the Petri dish.

To return to the role of Msb2 in filamentous growth, the phenotypes of loss of Msb2 and overexpression of Msb2, are shown in Figures 7 and 8. In the microscopic assay, *msb2* mutants do not show the elongated shape and do a poor job at unipolar budding. In

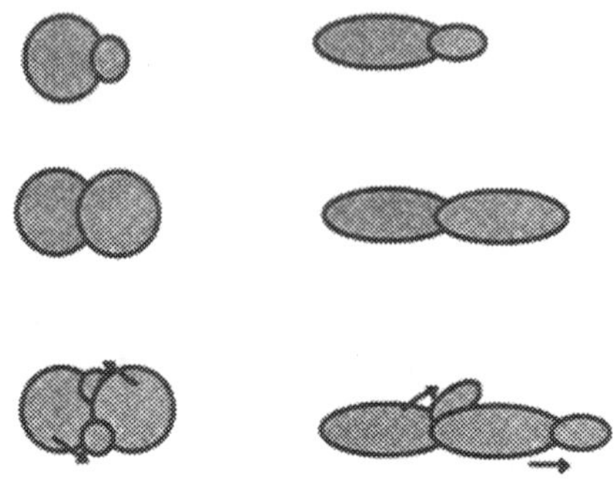

Figure 5. Yeast growth patterns. The budding pattern and cell shapes are contrasted for cells growing in abundant nutrients (axial, round) and in limiting nutrients (unipolar, elongated). This difference serves as a convenient microscopic assay for the switch from yeast form to filamentous form growth.

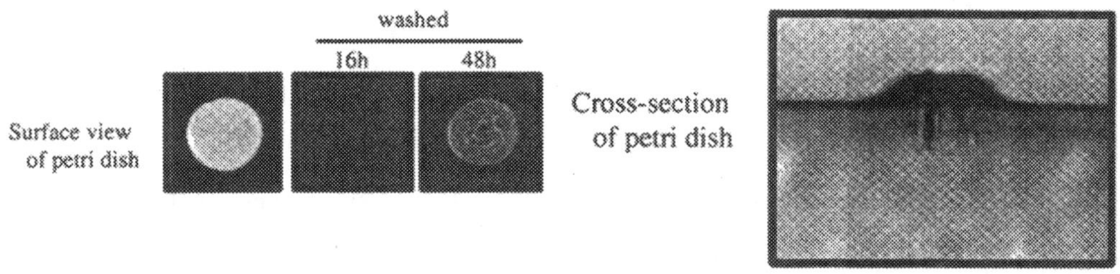

Figure 6. Plate-washing assay for filamentous growth. In the surface view of a Petri dish, a colony was allowed to grow for 48 hr and washed from the surface under running tap water. A scar that reflects the residue of the colony is readily visible. In cross-section, one can see that yeast cells have invaded the agar substratum.

contrast, overexpression of Msb2 leads to enhanced elongation and an exaggerated filamentous form. In the plate-washing assay, *msb2* mutants are substantially enfeebled for invasion of the agar.

Thus, we propose that the osmolarity and filamentous growth pathways are indeed distinct from each other at the head of the pathways. One pathway uses Sho1 alone and the other uses a combination of Sho1 and Msb2 (Figure 9). Each of these proteins is a presumptive transmembrane protein, so one can imagine that they are communicating in some way, perhaps by forming a heteromeric dimer. It is intriguing that Msb2 is not only a component of the pathway, but is also a transcription target of the pathway. Perhaps the increased transcription of the gene serves as a signal amplification or reinforcement mechanism (Figure 10).

Combinatorial Signaling: As described above, the filamentous growth pathway has unique components that distinguish it from the other signaling pathways that have been considered. Those unique components provide one means for a unique transcription readout from the pathway. However, the specificity of the overall filamentous growth pattern, as distinct from the mating response, for example, has at least one additional source. In particular, the filamentous growth pattern is the result of combinatorial activity of three distinct signal transduction pathways: the filamentous growth pathway that has been discussed thus far, and two additional pathways. This point of view emerged from

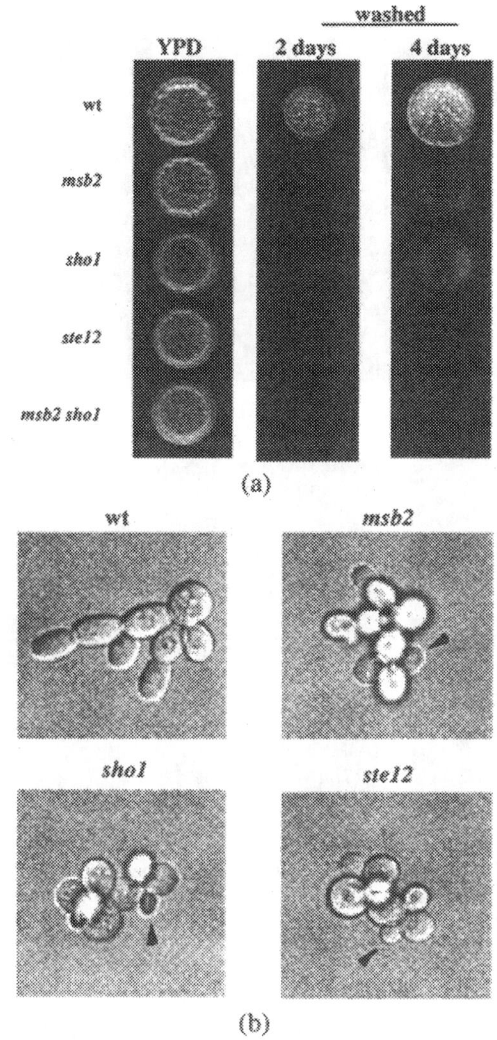

Figure 7. *MSB2* is required for filamentous growth. The results of plate-washing and microscopic assays are presented.

efforts to understand the control of filamentous growth by nutrient limitation. Our laboratory sought to determine whether a particular nutrient was key, simply by removing nutrients one at a time from the defined medium and asking whether filamentous growth was triggered (Cullen and Sprague, 2000). Removal of certain nutrients (e.g., amino acids), did not induce filamentous growth. However, removal of a fermentable carbon source led to vigorous filamentous growth (Figure 11). There is a well-known pathway, studied in Marian Carlson's lab, by which yeast cells sense whether glucose is present in the medium. This pathway includes a protein kinase called Snf1 and is distinct from the filamentous growth pathway (for review, see Carlson, 1999). We showed that *snf1* mutants do not invade the agar, or change their budding pattern and morphology. Hence, the Snf1 pathway is also required for the filamentous growth pattern.

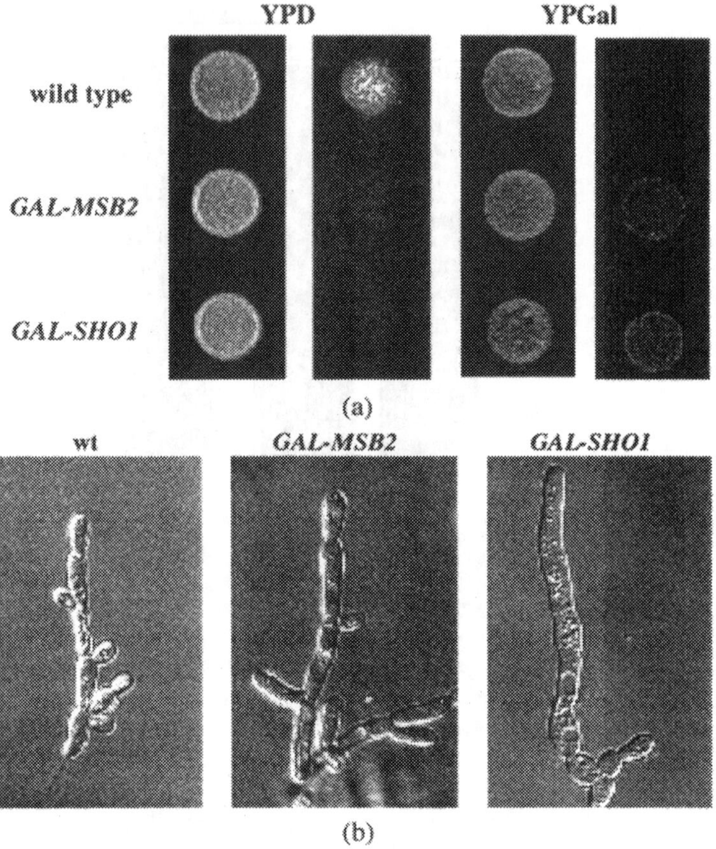

Figure 8. Overexpression of *MSB2* enhances filamentous growth. The *MSB2* gene was placed under control of the GAL1 promoter. In the plate-washing assay, the colony expressing Msb2 from this promoter shows a more vigorous invasion phenotype than does a colony not expressing the construct. Likewise, in the microscopic assay, the filaments formed are more exaggerated than are wild-type filaments.

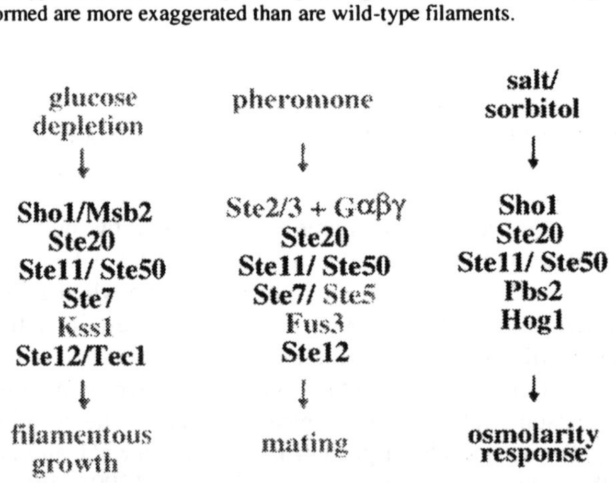

Figure 9. *MSB2* distinguishes the filamentous growth and osmolarity response pathways. The hypothesis that Msb2 forms a heteromeric "receptor" with Sho1 and that way distinguishes the top of the filamentous growth pathway from the top of the osmolarity response pathway is illustrated.

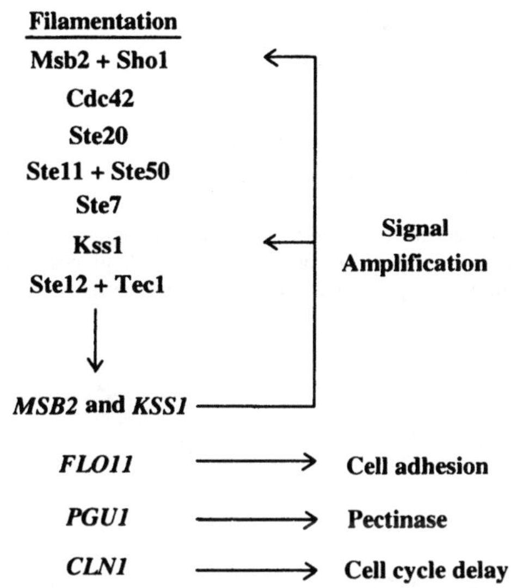

Figure 10. Signal amplification within the filamentous growth pathway. *MSB2* and *KSS1*, which encode components of the filamentous growth pathway, are themselves transcription targets of the pathway. This relationship may serve to amplify the initial signal generated by nutrient limitation. Three other transcription targets of the filamentous growth pathway are also noted. The products of these genes contribute to cell-cell adhesion (*FLO11*), agar invasion (*PGU1*), and cell elongation (*CLN1*).

Nutrient Removed	Cell Type
None	yeast form
Glucose	filamentous form
Sucrose	filamentous form
Mannose	filamentous form
Ammonia	yeast form
Amino acids	yeast form
Nucleotide	yeast form

Figure 11. The effect of removal of various nutrients on the growth habit of yeast cells.

How does Snf1 orchestrate the change in cell shape and the change in budding pattern (Figure 12)? Here, we will focus on the change in budding pattern. As a result of work from the Herskowitz lab and the Pringle lab, we know a reasonable amount about how yeast cells decide what budding pattern to undergo (Chant et al., 1995; Pringle et al., 1995; Herskowitz, 1997; Harkins et al., 2001). There is a set of proteins–and we will use Bud3 as an example–that is important in haploid cells for marking the axial site (Figure 13). A *bud3* mutant won't undergo axial budding. Similarly, there is a set of proteins—and we will use Bud8 as the example—that marks the distal pole, a pole that is normally used only in diploid cells. We asked whether Bud8 was important for the unipolar pattern that haploid cells undergo in response to glucose limitation. Indeed, Bud8 is required for invasive growth (Figure 14). In the plate-washing assay, there is very little invasion, and by microscopic assay *bud8* mutants exhibit axial—rather than unipolar—budding upon glucose deprivation.

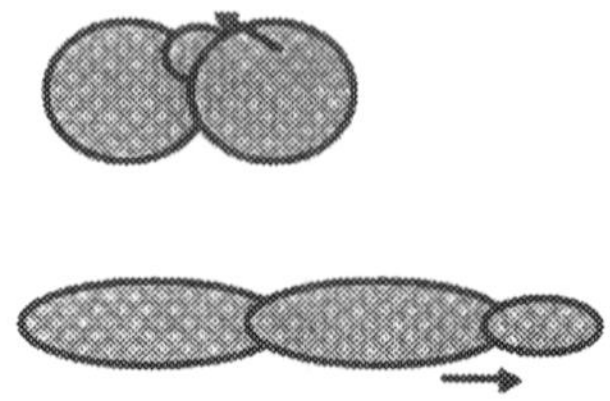

Figure 12. The effect of glucose on cell shape and budding pattern. In glucose-rich conditions, cells are spherical and the budding pattern is axial, as indicated by the arrow. In glucose-limited conditions, the cells are elongated and the budding pattern is unipolar, again as indicated by the arrow.

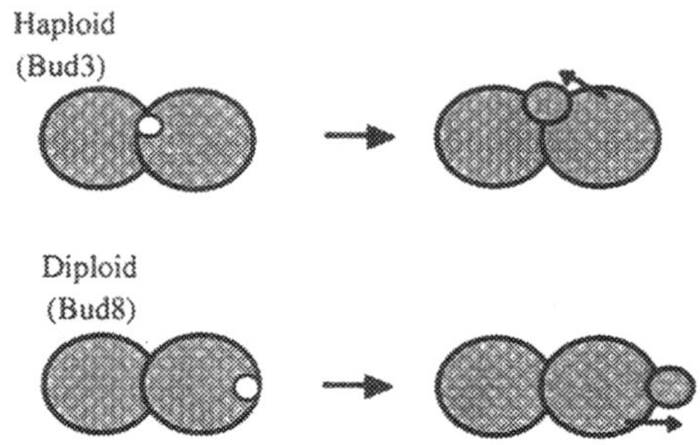

Figure 13. Bud site selection proteins direct bud emergence in different cell types. In haploid cells, Bud3 (and other proteins) mark the axial site and direct new budding to that site. In diploid cells, Bud8 marks the distal site and can direct budding from that position. The Bud8 protein is present and correctly localized in haploid cells, whether glucose is abundant or limiting, but the Bud8 mark is not used to direct bud emergence when glucose is abundant.

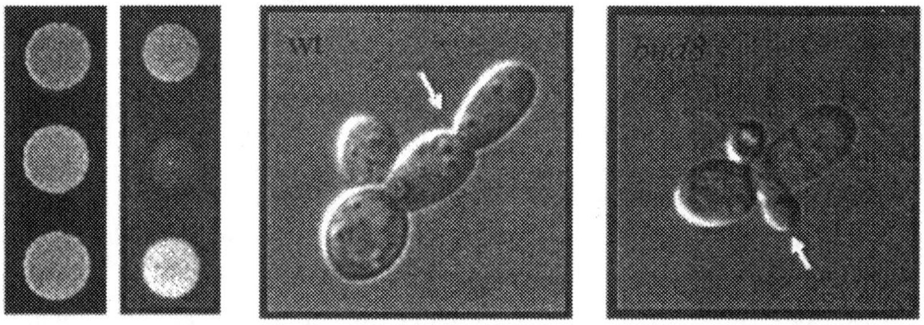

Figure 14. Bud8 is required for haploid invasive growth. *bud8* mutants are defective for agar invasion, as revealed by the plate-washing assay, and exhibit the axial budding pattern, as revealed by microscopic examination.

How is the glucose signal influencing the budding pattern? There is no change in the localization of known marks for budding pattern under glucose limitations; for example, Bud8 and Bud3 are properly localized under such conditions. However, glucose limitation has a dramatic effect on the abundance of another protein required for the axial pattern, Axl1. Axl1 is not a mark for the axial site but rather is thought in some way to process the axial site and make it competent for use. In glucose-rich conditions, the cells have abundant Axl1 protein but in glucose-deprived conditions, Axl1 protein disappears (Figure 15). Axl1 protein disappearance is a Snf1-dependent response (Cullen and Sprague, 2002).

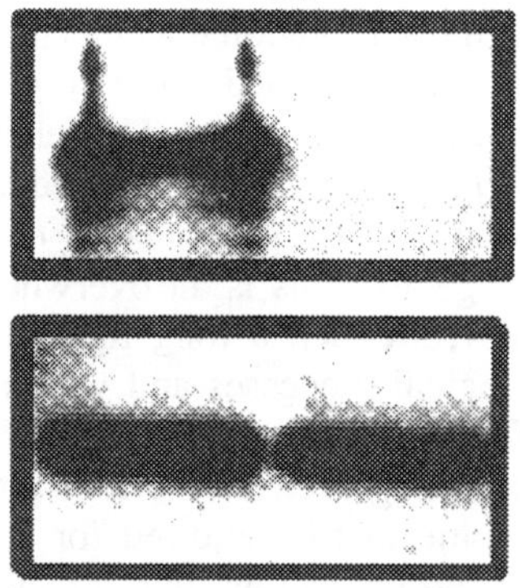

Figure 15. Axl1 protein abundance is regulated by glucose. Axl1, a protein required for the axial budding pattern, is abundant when cells are grown in the presence of glucose but is absent when glucose is limiting, as revealed by Western analysis using antibodies to Axl1 protein. Dpm1 serves as a loading control.

A third pathway, the Ras-cyclic AMP pathway, is also required for the overall filamentous growth response. The trigger that activates Ras remains elusive. Nonetheless, the observation that three distinct signal transduction pathways are required to orchestrate the overall filamentous growth response offers a new element of specificity. Filamentous growth is directed only when all three pathways are activated. Activation of only a single pathway leads to a different physiological response. Filamentous growth occurs by integrating three kinds of information: (i) that having to do with the Ste MAP kinase pathway activity, (ii) that having to do with glucose presence, and (iii) that having to do with Ras pathway activity.

8.3. STE20 INTERFACE WITH CELL BIOLOGY

Ste20 is the first protein kinase in the quartet of protein kinases that function in the pheromone response pathway, the filamentous growth pathway, and the osmolarity pathway. Might it do things in yeast cells besides participate in the signal transduction pathways? This question is motivated by experiments first done in Kim Nasmyth's lab (Cvrckova et al., 1995). Ste20 and a related protein kinase, Cla4, are both referred to as p21-activated kinases, or PAKs. They are regulated by Cdc42. Ste20, as has already been discussed, is involved in the forenamed signal transduction pathways and has also been argued to be involved in apical growth—the growth that happens at the very tip of the bud. Cla4 has also been argued to have a role in apical growth, to have a role in the G1 phase of the cell cycle, and a role in cytokinesis (Cvrckova et al., 1995; Benton et al., 1997; Holly and Blumer, 1999). As might be suspected, given these different roles, mutation of *STE20* or *CLA4* confers strikingly different phenotypes. However, a cell carrying mutations in both *STE20* and *CLA4* is dead. One interpretation of this finding is that Ste20 and Cla4 may share at least one activity and regulate the activity of an essential target. In the absence of one protein, the other can do the job. If this point of view is true, it implies that Ste20 has roles outside of the signal transduction pathways.

To learn what these roles might be, we sought new mutations that were lethal in a *cla4* mutant background. We used two different schemes to search for such synthetic lethal mutations (Mitchell and Sprague, 2001; Goehring et al., 2003). The first was based on random mutagenesis of the yeast genome and a colony-color assay to identify mutants (Figure 16). By this method, we identified about 10 genes. The second method was done in collaboration with Charlie Boone at the University of Toronto. In this effort, we crossed a strain carrying a *cla4* deletion to an ordered array of yeast mutants, individually deleted for each non-essential yeast gene (Tong et al., 2001). This effort identified roughly 70 synthetic lethal genes. This is an overwhelming number of genes. How can one begin to extract sense from such a long list? Many of the genes fall into groups involved in particular biological processes and we have chosen two of these groups to explore in some detail. The first group includes *BNI1, PEA2, BUD6*, and *SPA2*. This group was particularly interesting because the four proteins have been shown to form a complex called the polarisome that is required for apical growth, a process that Ste20 may influence. The other group caught our attention because single mutants were defective for filamentous growth, again a process that involves Ste20. Below, we discuss our findings with both groups.

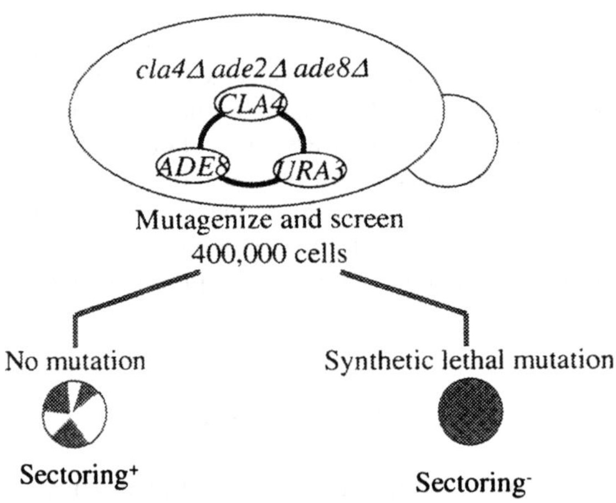

Complementation Group	Gene or ORF	# isolates
NCS1	*RRD1*	1
NCS2	YNL119w/ YNL120c	6
NCS3	*UBA4*	6
NCS4	*STE20*	2
NCS5	*BUD6*	2
NCS6	YGL211w	3
NCS7	N/D	2
NCS8	*BNI1*	5
NCS10	*ELP2*	2

Figure 16. Isolation of mutations that are synthetically lethal with *cla4*Δ. A colony sectoring assay was used to identify the desired mutants. An *ade2 ade8* strain forms white colonies, whereas an *ade2 ADE8* strain forms red colonies. An *ade2 ade8* strain carrying wild-type *ADE8* on the plasmid forms sectored colonies due to loss of the plasmid. The desired synthetic lethal mutations cause the formation of non-sectoring, red colonies because of loss of the *ADE8* plasmid, which also contains wild-type *CLA4*, cannot be tolerated.

Ste20 and the polarisome: Bni1, Pea2, Spa2 and Bud6 are components of the polarisome. Bni1 has homologs in many other species and has been argued, in some way, to regulate the actin cytoskeleton (Evangelista et al., 1997; Sheu et al., 1998). In yeast, Bni1 is known to interact with a large number of proteins and through those interactions has been implicated in many different aspects of cell biology. For example, Bni1 is known to bind a protein called Num1, which has a role in nuclear migration (Farkasovsky and Kuntzel, 2001). Of the presumptive functions of Bni1, is there a single one that is essential in the cell that lacks Cla4? The answer is satisfyingly simple. Loss of Pea2, Spa2 and Bud6, the other components of the polarisome, is lethal in a *cla4* background but loss of Num1–or other proteins–that are related to other Bni1 functions is not lethal. Hence, it seems reasonable to suppose that there is a specific connection between Ste20 and the polarisome. In particular, we wondered whether there was a connection between Ste20 and Bni1. We showed that Bni1 is a Ste20-dependent phosphoprotein, thereby making an explicit link between Ste20 and Bni1 (Goehring et al., 2003). Many questions remain, of course. Is phosphorylation of Bni1 by Ste20 direct? What sites are phosphorylated? What is the consequence of mutating those sites?

Ste20 and filamentous growth: Four of the synthetic lethal genes (*NCS* genes) are required for filamentous growth. Each of these genes is required for cell elongation during filamentous growth; however, none is required for unipolar budding. What might these genes encode? *URM1* and *NCS3 (UBA4)* are especially intriguing because they encode cousins of a ubiquitin modifying system (Furukawa et al., 2000). In particular, Urm1 has sequence similarity to ubiquitin and Ncs3 has sequence similarity to Uba1, the E1-conjugating enzyme of the classic ubiquitin pathway (Figure 17). We haven't identified versions of the ubiquitin pathway E2 and E3 enzymes that could serve to transfer the Urm1 moiety to ultimate target proteins but, nonetheless, we are quite interested to identify such targets of this hypothetical urmylation pathway, as we call it. As Marty Rechsteiner has already told you in his presentation to this symposium, there are other cousins to ubiquitin that are also present in yeast cells. Our interest in Urm1 is heightened further by the realization that there are strong homologs to this protein present in other species, including *Homo sapiens*. As a first step to identifying targets, we prepared antibodies to Urm1 and asked whether proteins that are tagged with Urm1 are visible in a yeast extract. Indeed, about a dozen proteins are readily visible in wild-type cells, but not visible in strains lacking either Urm1 or Ncs3. Hence, we suspect that there are a number of proteins that are modified by addition of the Urm1 moiety, and we have begun to purify these proteins so that they can be identified by mass spectrometry. It will be exciting to learn whether any of the urmylated proteins are themselves involved in filamentous growth in some way.

8.4. SUMMARY

Our studies have led to the realization that there are multiple mechanisms that control specificity of signal transduction pathway signaling and the attendant physiological response that ensues. Scaffolding proteins that organize signal transduction pathways are an especially powerful means to achieve specificity, but how general this mechanism is within yeast–and certainly beyond yeast–is an open question. Our studies have also started to reveal ways in which a protein, Ste20, first identified as a participant

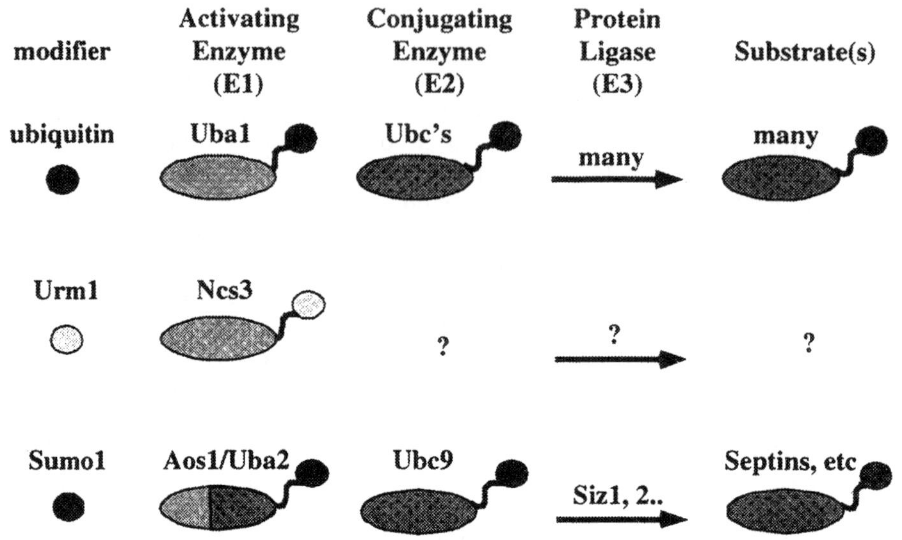

Figure 17. Ubiquitin and ubiquitin-like conjugation pathway. The classic ubiquitin pathway is shown with the identified E1 (Uba1), E2 (Ubc), and E3 enzymes illustrated. The hypothetical urmylation pathway, with the known modifier (Urm1) and E1 (Ncs3/Uba4) enzyme is illustrated. A second ubiquitin-like pathway, the sumo pathway, is also presented for contrast.

in signal transduction pathways, may also connect to the basic cell biology machinery. In essence, then, this protein serves as a branch point whereby the signal transduction pathway loses linearity and influences a number of different sorts of targets. Synthetic lethal genetic analysis has suggested that the polarisome and a new ubiquitin-like system may be targets of Ste20.

8.5. ACKNOWLEDGMENTS

We thank members of the lab, past and present, for stimulating conversations and insightful advice. This work was supported by research (GM30027) and training (GM07759) grants from the National Institutes of Health, and by a fellowship (AHA 120635Z) from the American Heart Association.

8.6. REFERENCES

Benton, B. K., Tinkelenberg, A., Gonzalez, I. and Cross, F. R., 1997, Cla4p, a *Saccharomyces cerevisiae* Cdc42p-activated kinase involved in cytokinesis, is activated at mitosis, *Mol. Cell Biol.* **17**:5067-5076.

Carlson, M., 1999, Glucose repression in yeast, *Curr. Opin. Microbiol.* **2**:202-207.

Chant, J., Mischke, M., Mitchell, E., Herskowitz, I. and Pringle, J. R., 1995, Role of Bud3p in producing the axial budding pattern of yeast, *J. Cell Biol.* **129**:767-778.

Cullen, P. J. and Sprague, G. F., Jr., 2000, Glucose depletion causes haploid invasive growth in yeast, *Proc. Natl. Acad. Sci. USA* **97**:13619-13624.

Cullen, P. J. and Sprague, G. F., Jr., 2002, The roles of bud-site-selection proteins during haploid invasive growth in yeast, *Mol. Biol. Cell* **13**:2990-3004.

Cvrckova, F., De Virgilio, C., Manser, E., Pringle, J. R. and Nasmyth, K., 1995, Ste20-like protein kinases are required for normal localization of cell growth and for cytokinesis in budding yeast, *Genes Dev.* **9**:1917-1830.

Evangelista, M., Blundell, K., Longtine, M. S., Chow, C. J., Adames, N., Pringle, J. R., Peter, M. and Boone, C., 1997, Bni1p, a yeast formin linking Cdc42p and the actin cytoskeleton during polarized morphogenesis, *Science* **276**:118-122.

Farkasovsky, M. and Kuntzel, H., 2001. Cortical Num1p interacts with the dynein intermediate chain Pac11p an cytoplasmic microtubules in budding yeast, *J. Cell Biol.* **152**:251-262.

Furukawa, K., Mizushina, N., Noda, T. and Ohsumi, Y., 2000, A protein conjugation system in yeast with homology to biosynthetic enzyme reaction of prokaryotes, *J. Biol. Chem.* **275**:7462-7465.

Gimeno, C. J., Ljungdahl, P. O., Styles, C. A. and Fink, G. R., 1992, Unipolar cell divisions in the yeast *S. cerevisiae* lead to filamentous growth; regulation by starvation and RAS, *Cell* **68**:1077-1090.

Goehring, A. S., Mitchell, D. A., Tong, A.H.Y., Keniry, M., Boone, C. and Sprague, G. F., Jr., 2003, Synthetic lethal analysis implicates Ste20p, a p21-activated protein kinase, in polarisome activation, *Mol. Biol. Cell*, in press.

Harkins, H. A., Page, N., Schenkman, L. R., De Virgilio, C., Shaw, S., Bussey, H. and Pringle, J. R., 2001, Bud8p and Bud9p, proteins that may mark the sites for bipolar budding in yeast, *Mol. Biol. Cell* **12**:2497-2518.

Herskowitz, I., 1997, Building organs and organisms: elements of morphogenesis exhibited by budding yeast, *Cold Spring Harb. Symp. Quant. Biol.* **62**:57-63.

Holly, S. P. and Blumer, K. J., 1999, PAK-family kinases regulate cell and actin polarization throughout the cell cycle of *Saccharomyces cerevisiae*, *J. Cell Biol.* **147**:845-856.

Mitchell, D. A. and Sprague, G. F., 2001, The phosphotyrosyl phosphatase activator, Ncs1p (Rrd1p), functions with Cla4p to regulate the G(2)/M transition in *Saccharomyces cerevisiae*, *Mol. Cell. Biol.* **21**:488-500.

O'Rourke, S. M. and Herskowitz, I., 1998, The Hog1p MAPK prevents cross talk between the HOG and pheromone response MAPK pathways in *Saccharomyces cerevisiae*, *Genes Dev.* **12**:2874-2886.

Posas, F. and Saito, H., 1997, Osmotic activation of the HOG MAPK pathway via Ste11p MAPKKK: scaffold roles of Pbs2p MAPKK, *Science* **276**:1702-1705.

Pringle, J. R., Bi, E., Harkins, H. A., Zahner, J. E., De Virgilio, C., Chant, J., Corrado, K. and Fares, H., 1995, Establishment of cell polarity in yeast, *Cold Spring Harb. Symp. Quant. Biol.* **60**:729-744.

Sheu, Y. J., Santos, B., Fortin, N., Costigan, C. and Snyder, M., 1998, Spa2p interacts with cell polarity proteins and signaling components involved in yeast cell morphogenesis, *Mol. Cell. Biol.* **18**:4053-4069.

Tong, A. H. Evangelista, M., Parsons, A. B., Xu, H., Bader, G. D., Page, N., Robinson, M., Raghibizadeh, S., Hogue, C. W., Bussey, H., et al., 2001, Systematic genetic analysis with ordered arrays of yeast deletion mutants, *Science* **294**:2364-2368.

Yashar, B., Irie, K., Printen, J. A., Stevenson, B. J., Sprague, G. F., Jr., Matsumoto, K. and Errede, B., 1995, Yeast MEK-dependent signal transduction: Response thresholds and parameters affecting fidelity, *Mol. Cell. Biol.* **15**:6545-6553.

INDEX

MIX
Papier aus verantwortungsvollen Quellen
Paper from responsible sources
FSC® C105338

If you have any concerns about our products,
you can contact us on
ProductSafety@springernature.com

In case Publisher is established outside the EU,
the EU authorized representative is:
Springer Nature Customer Service Center GmbH
Europaplatz 3, 69115 Heidelberg, Germany

Printed by Libri Plureos GmbH
in Hamburg, Germany